# Graph Attack!

## UNDERSTANDING CHARTS AND GRAPHS

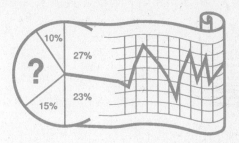

This book is a companion volume to

*Map Attack!*
Understanding Globes and Maps

# Graph Attack!

## UNDERSTANDING CHARTS AND GRAPHS

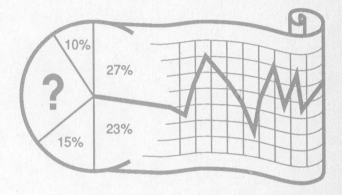

# JACK WARNER

**CAMBRIDGE** Adult Education
REGENTS/PRENTICE HALL,
Englewood Cliffs, NJ 07632

Executive editor: James W. Brown
Editorial supervision: John Chapman
Production supervision and interior design: Noël
    Vreeland Carter
Cover design: Bruce Kenselaar
Cover art: Garry Gay, The Image Bank
Technical illustrations: Networkgraphics
Art illustration: Don Martinetti, D.M. Graphics, Inc.
Pre-press buyer: Ray Keating
Manufacturing buyer: Lori Bulwin
Scheduler: Leslie Coward

© 1993 by REGENTS/PRENTICE HALL
A Division of Simon & Schuster
Englewood Cliffs, New Jersey 07632

Printed in the United States of America
10 9 8 7 6

ISBN 0-13-363110-9

Prentice-Hall International (UK) Limited, *London*
Prentice-Hall of Australia Pty. Limited, *Sydney*
Prentice-Hall of Canada Inc. *Toronto*
Prentice-Hall Hispanoamericana, S.A., *Mexico*
Prentice-Hall of India Private Limited, *New Delhi*
Prentice-Hall of Japan, Inc. *Tokyo*
Simon & Schuster Asia Pte. Ltd., *Singapore*
Editora Prentice-Hall do Brasil, Ltda., *Rio de Janeiro*

# Graph Attack!

## UNDERSTANDING CHARTS AND GRAPHS

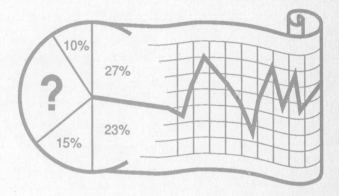

## JACK WARNER

**CAMBRIDGE** Adult Education
REGENTS/PRENTICE HALL,
Englewood Cliffs, NJ 07632

Executive editor: James W. Brown
Editorial supervision: John Chapman
Production supervision and interior design: Noël
    Vreeland Carter
Cover design: Bruce Kenselaar
Cover art: Garry Gay, The Image Bank
Technical illustrations: Networkgraphics
Art illustration: Don Martinetti, D.M. Graphics, Inc.
Pre-press buyer: Ray Keating
Manufacturing buyer: Lori Bulwin
Scheduler: Leslie Coward

Printed in the United States of America
10  9  8  7  6


ISBN 0-13-363110-9


Prentice-Hall International (UK) Limited, *London*
Prentice-Hall of Australia Pty. Limited, *Sydney*
Prentice-Hall of Canada Inc. *Toronto*
Prentice-Hall Hispanoamericana, S.A., *Mexico*
Prentice-Hall of India Private Limited, *New Delhi*
Prentice-Hall of Japan, Inc. *Tokyo*
Simon & Schuster Asia Pte. Ltd., *Singapore*
Editora Prentice-Hall do Brasil, Ltda., *Rio de Janeiro*

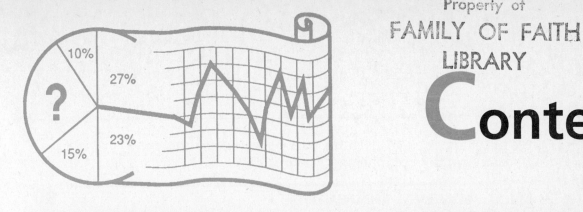

# Contents

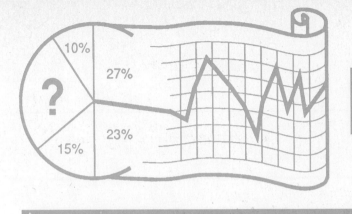

# Introduction

This book provides instruction and practice in working with various types of tables, charts and graphs. It uses factual information from a wide variety of sources to help users learn how graphs are put together, as well as how information can be obtained from them. The subject matter of each chapter is the kind adults encounter both in daily living and on such tests as the GED.

The format of the chapters in this book is consistent throughout. Each chapter carefully integrates reading, writing and thinking skills. Each chapter opens with How Much Do You Already Know?— four brief multiple-choice questions that serve as a preview of the chapter's content. Each chapter is divided into several manageable segments. Short-answer questions throughout the instruction allow students to apply concepts immediately.

There is a Warmup at the end of each instructional segment which serves as a brief summary quiz for that section. How Carefully Did You Read?—an end-of-chapter exercise made up of both multiple-choice and short-answer questions—allows students to review and apply all the concepts covered in a chapter.

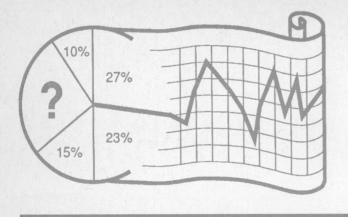

## How Much Do You Already Know?

**Choose a correct completion for each statement. If you are not sure about an answer, do not guess.**

1. Most tables present their information in

   ☐ a. lists of words.

   ☐ b. rows and columns of numbers.

   ☐ c. dollars and cents.

2. The title of a table tells you

   ☐ a. what information the table contains.

   ☐ b. how the information is presented.

   ☐ c. how recent the information is.

3. The source line below a table tells you

   ☐ a. where the author found the table.

   ☐ b. where the information in the table came from.

   ☐ c. what type of information the table contains.

4. Tables present many kinds of information including

   ☐ a. population figures.

   ☐ b. reasons for changes in population figures.

   ☐ c. suggestions for finding further information on a topic.

**Check your answers on page 147.**

# Tables

**Tables** provide us with an effective way of looking at groups of related numbers. They present us with many specific pieces of information in a form that makes it easy for us to compare those bits of data with each other.

## PARTS OF A TABLE

A table is a display of information, usually given in numbers. It is arranged in some orderly fashion—usually in **columns** and **rows**. Tables can present many different kinds of information including population figures, average temperatures, and workers' salaries. Look at the table in Figure 1.1.

Column → Attendance at Adult Education Classes
Johnson City
October 1992 } Title

| DAY | WEEK 1 | WEEK 2 | WEEK 3 | WEEK 4 |
|-----|--------|--------|--------|--------|
| MON | 198 | 208 | 202 | 197 | ← Row
| TUE | 207 | 210 | 200 | 216 |
| WED | 212 | 219 | 214 | 211 |
| THU | 217 | 212 | 205 | 209 |
| FRI | 193 | 196 | 188 | 203 |

Source ⟶ Board of Education.

**Figure 1.1  The Parts of a Table**

The topic of a table is always found in the **title** which usually appears at the top. What kind of information does this table show? _____

_____

The table shows attendance at adult education classes in Johnson City for October 1992. Notice that the information is arranged in columns and rows. Columns run from top to bottom. The first column in Figure 1.1 contains the **row headings** for the table. What is the heading of that column? _____

The word *day* appears at the head of the first column.

 What are the headings of the other four columns in the table?

_____

_____

_____

_____

*Week 1, Week 2, Week 3*, and *Week 4* are the other four column headings. The numbers in these columns show daily attendance for each of the weeks.

 Rows are read from left to right. What kind of information is listed in each row?

_____

_____

Each row tells how many students attended classes on each day of a given week. A line below a table usually tells the **source** of the information contained in the table.

 What is the source of the information in the table in Figure 1.1? _____

_____

The information comes from the Board of Education.

Always notice the source when you read a table. It may suggest how accurate and reliable the information in the table is.

## READING A TABLE

Suppose you wanted to find out how many students attended classes on Thursday in Week 3. First, look for Thursday in the row headings. Then trace across the Thursday row until you come to the Week 3 column. Where the row and column meet you will find how many students attended classes—205.

 How many students attended classes on Wednesday in Week 4? _____

On Wednesday in Week 4, 211 students attended classes.

## USING A TABLE

Sometimes it will be useful to add or subtract numbers on a table in order to learn something not shown there directly.

Suppose you want to know how many more students attended class on Thursday of Week 1 than on Friday of that week. First, read down the Week 1 column until you come to the Thursday row.

 How many students attended class on that Thursday? _____

According to the table, 217 students attended class that day.

 After that read down the Week 1 column to the Friday row. How many students attended class on that Friday? _____

The table shows that 193 students attended class that day.

Now subtract the Friday total from the Thursday total. How many more students attended class on Thursday of Week 1 than on Friday of that week?

_____

Twenty-four more students came to class on Thursday of Week 1 than on Friday.

**WARMUP**

**Use the table in Figure 1.1 to answer these questions.**

1. How many students attended classes on Wednesday in Week 1? _____

2. On what day of that week did the fewest students attend classes? _____

3. On which day in which week did the most students attend classes? _____

4. What was the total number of students to attend classes in Week 4? _____

5. How many more students attended classes on Thursday in Week 2 than attended classes on Friday in that same week? _____

**Check your answers on page 147.**

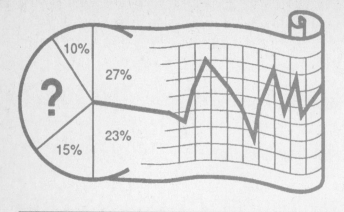

## How Carefully Did You Read?

**A. Choose the correct completion for each statement.**

1. The kind of information a table shows can be found in its

   ☐ a. source.
   ☐ b. column headings.
   ☐ c. title.

2. The source of a table tells

   ☐ a. what information is in the table.
   ☐ b. where the information in the table came from.
   ☐ c. why the information in the table is important.

3. In a table, columns run from

   ☐ a. top to bottom.
   ☐ b. left to right.
   ☐ c. right to left.

4. A table is a display of information, usually in

   ☐ a. pictures.
   ☐ b. words.
   ☐ c. numbers.

5. In a table, rows run from

☐ a. left to right
☐ b. bottom to top.
☐ c. right to left.

6. You can learn something not shown directly on a table by

☐ a. adding or subtracting numbers on the table.
☐ b. changing the title of the table.
☐ c. changing the column headings.

**B. Use the table in Figure 1.2 to complete the sentences.**

Sales of Single Family Homes
Selected Rhode Island Cities

| CITY | 1986 | 1987 | 1988 | 1989 | 1990 |
|------|------|------|------|------|------|
| MIDDLETOWN | 237 | 301 | 310 | 303 | 275 |
| NEWPORT | 481 | 374 | 371 | 329 | 149 |
| PORTSMOUTH | 612 | 574 | 550 | 416 | 211 |
| WARWICK | 783 | 694 | 649 | 512 | 193 |

Source: Rhode Island Board of Realtors.

Figure 1.2  Sales of Single Family Homes

1. The title of the table is _____.

2. In Middletown, _____ houses were sold in 1988.

3. The total number of houses sold in Warwick during the years shown in the table is

_____.

4. The city with the fewest house sales in 1987 is _____.

5. The city with the fewest house sales in 1990 is _____.

6. According to the table, the year in which the greatest number of houses was sold in

any one city was _____.

7. The source for the information in the table is _____.

8. The difference in the number of houses sold in Newport in 1988 and in 1990 amounted

to _____ houses.

**Check your answers on page 147.**

Look back at **How Much Do You Already Know?** on page 2. Did you complete each statement correctly? If not, can you do so now?

TITLE:

| YEAR | POPULATION |
|------|------------|
|      |            |
|      |            |
|      |            |
|      |            |
|      |            |
|      |            |
|      |            |
|      |            |

**Figure 1.3  Completing a Table**

ARE YOU READY FOR THE CHALLENGE of making a table from information given you? If you think you are, use the data that follow to fill in the table in Figure 1.3. Then write a title for the table and add a source line below the table.

The U.S. Census Bureau is the government office that counts how many people live in the United States. According to its figures, the population of the United States in 1790 was 3,900,000. In 1800, the population had risen to 5,300,000; by 1810 it was 7,200,000. By 1820 there were 9,600,000 people living in the country. Ten years later, in 1830, the population was 12,800,000. In 1840 it was 17,000,000, and in 1850, 23,000,000. By 1860, the year before the American Civil War began, the population reached 31,000,000 people.

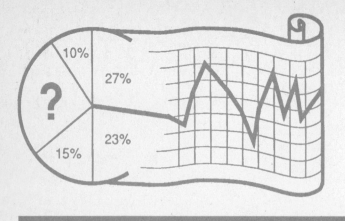

## How Much Do You Already Know?

**Choose the correct completion for each statement. If you are not sure about an answer, do not guess.**

1. Charts present information

   ☐ a.  using only rows and no columns.

   ☐ b.  in numbers, symbols or pictures.

   ☐ c.  without a title.

2. One feature found on some charts but not on tables is

   ☐ a.  a title.

   ☐ b.  a source.

   ☐ c.  a key.

3. You would find ♙ = 2 people in the

   ☐ a.  title of a chart.

   ☐ b.  key of a chart.

   ☐ c.  source of a chart.

4. On a chart the rows run

   ☐ a.  from left to right.

   ☐ b.  from right to left.

   ☐ c.  from top to bottom.

**Check your answers on page 147.**

# Charts

## PARTS OF A CHART

Like tables, **charts** present information in **rows** and **columns**. But whereas tables nearly always use numbers, charts may present information in symbols or pictures as well. Look at the chart in Figure 2.1.

Temporary Workers Hired
1992
Video City

| TIME PERIOD | DEPARTMENTS | | |
| --- | --- | --- | --- |
| | SALES | DELIVERIES | CUSTOMER SERVICE |
| JAN–MAR | | | |
| APR–JUN | | | |
| JUL–SEP | | | |
| OCT–DEC | | | |

Source: Human Resources.

Key (  = 2 People

**Figure 2.1   The Parts of a Chart**

According to its **title**, what kind of information does the chart in Figure 2.1 show? _____

This chart is a record of temporary workers hired in one year at a store called Video City.

The chart shows the number of temporary workers hired in each of three departments. What are those departments? _____

_____

According to the chart, Video City hired temporary workers in three departments: Sales, Deliveries, and Customer Service.

The names of the three departments also act as **column headings**. As in tables, columns in a chart are read from top to bottom.

 Headings for the rows in the chart are listed on the left side of the chart. What do these **row headings** show? _____

The row headings tell when the temporary workers were hired. The workers were hired in one of four time periods: January to March, April to June, July to September, or October to December.

Notice that **symbols** instead of numbers are used to show the number of people hired in any one time period. To learn what the symbol ᛁ stands for, you must look for the **key**. In this chart the key appears at the bottom.

What does the symbol ᛁ stand for? _____

Each ᛁ on the chart represents 2 people. The chart has ᛁᛁᛁᛁ symbols in one section. This represents 8 people.

## READING A CHART

Suppose you want to learn how many temporary workers were hired by the Customer Service Department during the period July through September. First, read down the column until you reach the abbreviation JULY–SEPTEMBER. Then trace across this row until you find the heading *Customer Service*. The place where the column and row meet is called a **cell**.

How many temporary workers were hired in that time period? _____

According to the chart, 10 temporary workers were hired by Customer Service in the July to September time period. (Each ᛁ represents 2 people, and there are 5 of these symbols in that cell.)

## Use the chart in Figure 2.1 to answer these questions.

1. How many workers were hired to make deliveries during the period April through June?_____

2. In all, how many temporary workers did Video City hire during 1992? _____

3. How many temporary workers did Video City hire in all three departments during the period July through September? _____

4. During which period did Video City hire the fewest temporary workers? _____

5. How many cells are there in the Video City chart? _____

**Check your answers on page 147.**

## CHARTS THAT USE NUMBERS

Of course not all charts use symbols to present information. Some, like the chart in Figure 2.2, use numbers.

What kind of information does the chart in Figure 2.2 show? _____

_____

Mileage Between Principal Cities
Nevada
(Distances in Miles)

| | BOULDER CITY | CARSON CITY | ELKO | ELY | FALLON | HAWTHORNE | LAS VEGAS | LOVELOCK | RENO |
|---|---|---|---|---|---|---|---|---|---|
| CARSON CITY | 461 | | 306 | 319 | 62 | 127 | 439 | 109 | 32 |
| ELKO | 481 | 306 | | 190 | 253 | 312 | 471 | 197 | 289 |
| ELY | 303 | 319 | 190 | | 257 | 271 | 281 | 313 | 318 |
| LAS VEGAS | 22 | 439 | 471 | 281 | 384 | 312 | | 440 | 450 |
| RENO | 472 | 32 | 289 | 318 | 61 | 138 | 450 | 92 | |

Department of Transportation.

**Figure 2.2  A Mileage Chart**

The chart in Figure 2.2 shows the mileage between several cities in the state of Nevada.

What is the source for the information in this chart? _____

_____

The information in this chart comes from the Department of Transportation (D.O.T.).

## READING A MILEAGE CHART

Assume that you are about to drive from Reno to Fallon and you want to know how many miles the trip will be. To find the mileage from the chart, first find the city you are starting from. Look down the headings row at the left of the chart for Reno. Then trace across that row until you find the cell under Fallon, the city you are traveling to. The number in that box shows the distance between those two cities.

 What is the mileage between Reno and Fallon? _____

According to the mileage chart, the distance between those two cities is 61 miles.

## WARMUP B

**Use the chart in Figure 2.2 to answer these questions.**

1. What is the distance between Carson City and Reno? _____

2. What is the distance between Ely and Lovelock? _____

3. In driving from Elko to Reno, and then from Reno to Ely, how many miles would you travel in all? (You must add to find the answer.) _____

4. How much closer to each other are Carson City and Fallon than Carson City and Boulder City? _____

5. According to the chart, which two cities are farthest apart? _____

**Check your answers on page 147.**

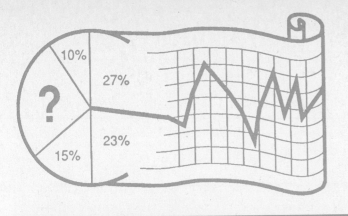

## How Carefully Did You Read?

**A. Choose the correct completion for each statement.**

1. The title of a chart tells

   ☐ a. where the information in that chart came from.

   ☐ b. the meaning of any symbols or pictures in the chart.

   ☐ c. what information the chart shows.

2. On a chart the key explains

   ☐ a. the meaning of any symbols in the chart.

   ☐ b. where the information in the chart came from.

   ☐ c. the date the information in the chart was gathered.

3. Charts present information

   ☐ a. only in numbers.

   ☐ b. only in pictures or symbols.

   ☐ c. in numbers or in pictures and symbols.

4. One feature found on charts but not on tables is a

   ☐ a. title.

   ☐ b. key.

   ☐ c. source.

5. As in tables, columns in charts are read from

☐ a. left to right.
☐ b. top to bottom.
☐ c. right to left.

6. The place where a column meets a row is called a

☐ a. cell.
☐ b. list.
☐ c. chart.

**B. Use the chart in Figure 2.3 to complete each statement.**

1. The title of the chart is _____.

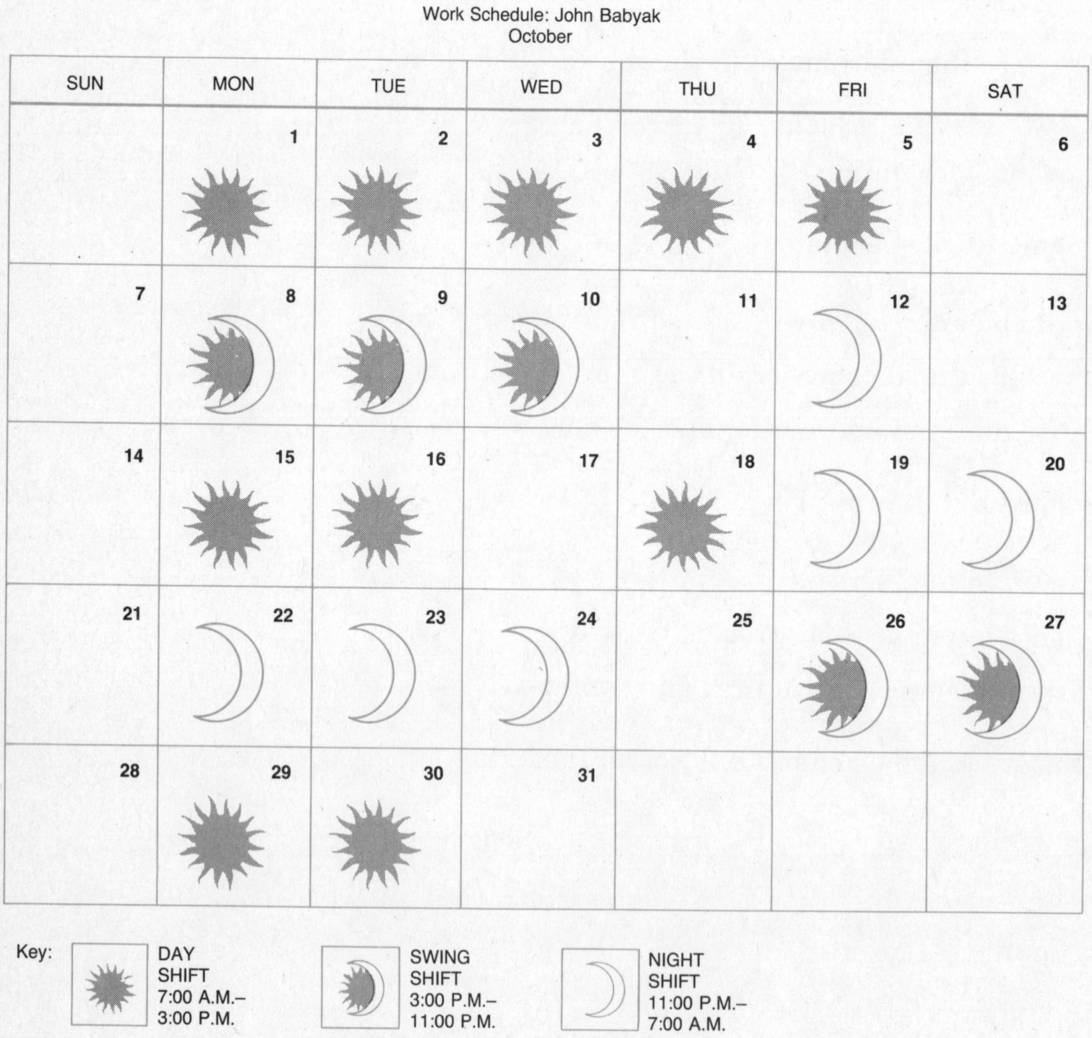

Work Schedule: John Babyak
October

| SUN | MON | TUE | WED | THU | FRI | SAT |
|---|---|---|---|---|---|---|
| | 1 ☀ | 2 ☀ | 3 ☀ | 4 ☀ | 5 ☀ | 6 |
| 7 | 8 ☀☾ | 9 ☀☾ | 10 ☀☾ | 11 | 12 ☾ | 13 |
| 14 | 15 ☀ | 16 ☀ | 17 | 18 ☀ | 19 ☾ | 20 ☾ |
| 21 | 22 ☾ | 23 ☾ | 24 ☾ | 25 | 26 ☀☾ | 27 ☀☾ |
| 28 | 29 ☀ | 30 ☀ | 31 | | | |

Key: ☀ DAY SHIFT 7:00 A.M.– 3:00 P.M.    ☀☾ SWING SHIFT 3:00 P.M.– 11:00 P.M.    ☾ NIGHT SHIFT 11:00 P.M.– 7:00 A.M.

**Figure 2.3  A Work Schedule Chart**

2. In all, John Babyak will work _____ of the 31 days in October.

3. During October, John Babyak will work _____ day shifts.

4. The number of swing shifts he will work is _____.

5. John Babyak has a total of _____ days off during October.

6. The one day of the week that John Babyak does not work is _____.

7. The last day shift John Babyak will work falls on _____.

8. The one day of the week on which John Babyak will work only day shifts is _____.

**Check your answers on page 147.**

Look back at How Much Do You Already Know? on page 10. Did you complete each statement correctly? If not, can you do so now?

TITLE: August

| SUN | MON | TUE | WED | THU | FRI | SAT |
|---|---|---|---|---|---|---|
| | | | 1<br><br>Tidewater | 2<br><br>Tidewater | 3<br><br>Tidewater | 4<br><br>Buffalo |
| 5<br><br>Buffalo | 6 | 7 | 8 | 9 | 10 | 11 |
| 12 | 13 | 14 | 15 | 16 | 17 | 18 |
| 19 | 20 | 21 | 22 | 23 | 24 | 25 |
| 26 | 27 | 28 | 29 | 30 | 31 | |

Key: ▨ Home Game  ▧ Away Game

**Figure 2.4  Completing a Chart**

ARE YOU READY FOR THE CHALLENGE of making a chart from information given you? If you think you are, use the data that follow to fill in the chart in Figure 2.4. Be sure to add a title. Fill in the location of each game and shade the home game squares in gray. The first few parts of the chart have been done for you.

Here is the schedule of baseball games, both home and away games, for the Pawtucket Red Sox during August.

August 1, 2, 3: Tidewater, at home
August 4, 5: Buffalo, away
August 6, 7, 8: Louisville: Away
August 9, 10, 11: Nashville, away
August 12, 13, 14: Indianapolis, away
August 15, 16, 17: Richmond, at home
August 18, 19, 20: Rochester, at home
August 21, 22, 23: Scranton, at home
August 24, 25, 26: Richmond, away
August 27, 28, 29: Tidewater, away
August 30, 31: Rochester, at home

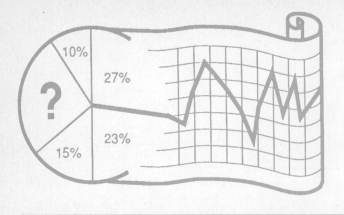

## How Much Do You Already Know?

**Choose the correct completion for each statement. If you are not sure about an answer, do not guess.**

1. The thick lines, or bars, in a bar graph are drawn

   ☐ a. only vertically.
   ☐ b. only horizontally.
   ☐ c. either vertically or horizontally.

2. The numbers running along the bottom of a graph are called

   ☐ a. the source.
   ☐ b. the horizontal scale.
   ☐ c. the vertical scale.

3. Bar graphs always have

   ☐ a. a source.
   ☐ b. two scales.
   ☐ c. a key.

4. On a bar graph, the longer the bar

   ☐ a. the lower its value.
   ☐ b. the greater its value.
   ☐ c. the more accurate the graph is.

**Check your answers on page 148.**

# Bar Graphs

Bar graphs use thick lines or bars to compare sets of figures. Figure 3.1 shows a **vertical bar graph** in which the bars run up and down.

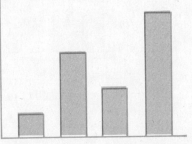

Figure 3.1   A Vertical Bar Graph

Figure 3.2 shows a **horizontal bar graph** in which the bars run from side to side.

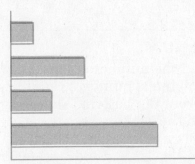

Figure 3.2   A Horizontal Bar Graph

Whether a vertical or a horizontal bar graph is used to compare sets of figures is really up to the person making the graph. However, if the information compares, say, the height of buildings, or a growth in sales over a period of time, a vertical bar graph will likely be used. On the other hand, if the figures compare, say, the lengths of rivers, a horizontal bar graph may be the appropriate choice.

Bar graphs are often used to show trends—that is, upward and downward movements—over a period of time.

## VERTICAL BAR GRAPHS

 Look at the bar graph in Figure 3.3. Is this a vertical or a horizontal bar graph?

_____

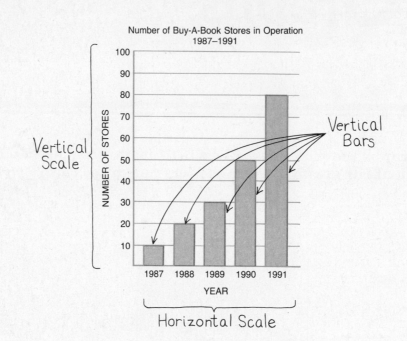

**Figure 3.3    The Parts of a Bar Graph**

Since the bars go up and down, Figure 3.3 is an example of a vertical bar graph.

 What kind of information does the graph show?_____

_____

The graph shows the increase in the number of Buy-A-Book stores from 1987 to 1991. You will notice that there is a column of numbers running down the left side of the graph, and a row of numbers running across the bottom. These are called **scales**. The figures on the left side of the graph are called the **vertical scale**.

 What does the vertical scale on the graph show? _____

_____

It shows the number of stores in the Buy-A-Book chain.
The numbers running across the bottom of the graph are called the **horizontal scale**.

 What does the horizontal scale on the graph show? _____

_____

It represents the years from 1987 through 1991.

## READING A VERTICAL BAR GRAPH

Suppose you wanted to find out how many stores the Buy-A-Book company had in 1989. First, find the bar that represents 1989. Then trace a line across the graph from the top of that bar to the vertical scale on the left.

 How many stores were operating in 1989? _____

According to the graph, Buy-A-Book had 30 stores in 1989.

## WARMUP A

**Use the vertical bar graph in Figure 3.3 to answer these questions.**

1. How many stores did Buy-A-Book operate in 1987?_____

2. How many stores did the company operate in 1988?_____

3. What was the *increase* in the number of stores between 1987 and 1988?

   _____

4. How many new stores did Buy-A-Book add between 1987 and 1990?_____

5. Does the graph show an increasing or decreasing trend in the number of Buy-A-Book

   stores?_____

**Check your answers on page 148.**

## HORIZONTAL BAR GRAPHS

Figure 3.4 shows a horizontal bar graph. What any bar graph shows is always found in the title, which most often appears at the top.

 What is the title of this bar graph? _____

The title is "Yearly Consumption of Certain Foods in the United States."

 Like a vertical bar graph, a horizontal bar graph has two scales. What does the

vertical scale at the left of the graph show? _____

The vertical scale shows the kinds of food consumed (eaten).

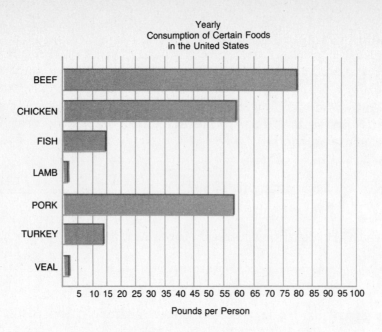

Figure 3.4  Yearly Consumption of Certain Foods

  The second scale runs across the bottom of the graph. What does this horizontal

scale show? _____

This scale shows the amount of food consumed in pounds per person in a given year. In other words, a bar which reaches a 10 on the horizontal scale indicates that each person in the United States ate an average of 10 pounds of that food in that particular year.

## READING A HORIZONTAL BAR GRAPH

As in vertical bar graphs, the length of a bar shows its value. The longer the bar, the greater the value.

Suppose you wanted to find out the average number of pounds of fish each person in the United States eats in a year. First, look down the left scale until you come to the bar that represents fish consumption. Locate the end of the bar and then trace down until to reach the horizontal scale at the bottom of the graph.

  How many pounds of fish does each person eat?_____

According to the graph, each person in the United States eats 15 pounds of fish a year.

## ESTIMATING AMOUNTS ON A BAR GRAPH

If you look closely at the graph, you will see that most of the bars end somewhere between one measure and the next on the horizontal scale. When this happens, you must estimate, or judge, the actual measure.

For example, locate the end of the bar representing the amount of chicken consumed in a year and trace a line down to the horizontal scale. The bar ends between 55 and 60 pounds

on the bottom scale. But it is much closer to 60 pounds than to 55 pounds. A good estimate, then, might be 59 pounds as the average amount of chicken a person eats in one year.

Graphs like this one are useful in comparing the average amount of food consumed in a given year with the amount of that same food that you eat in a year. You will likely find differences. Perhaps you eat less beef and more lamb, for example. Why are there differences? The reason is that the graph shows an *average* consumption of each food. To find this average, the total number of pounds consumed of each food was divided by the total number of people in the United States. The results are shown on the graph.

**WARMUP B**

**Use the horizontal bar graph in Figure 3.4 to answer these questions.**

1. Which food do Americans consume the most of each year? _____

2. Which food do Americans consume the least of each year? _____

3. About how many pounds of pork do Americans eat in a year? _____

4. About how many pounds of chicken and turkey combined do Americans eat in a year?

   _____

5. About how much more chicken than fish do Americans eat in a year? _____

**Check your answers on page 148.**

**EXTENDED BAR GRAPHS**

In some bar graphs the information given in a single bar is broken down into two or more segments. These are called **extended bar graphs** or **stacked bar graphs**. The bars on an extended bar graph may run either vertically or horizontally.

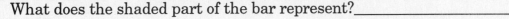 Figure 3.5 is an extended bar graph. What kind of information does it show?

_____

The graph shows the number of male and female workers at the Diaz Messenger Service between the years 1970 and 2000.

Notice that the scale representing the year 2000 has an asterisk (*) following it. Look at the bottom of the graph for the footnote which explains that this bar contains estimated figures. This estimate represents a projection, or goal, for the Diaz company. It tells the number of employees the company *hopes* to have by the year 2000.

Notice also that each bar is divided into two parts. One part of the bar is shaded, while the rest is not. To learn what the parts of a bar mean, look at the key at the bottom of the graph.

What does the shaded part of the bar represent?_____

_____

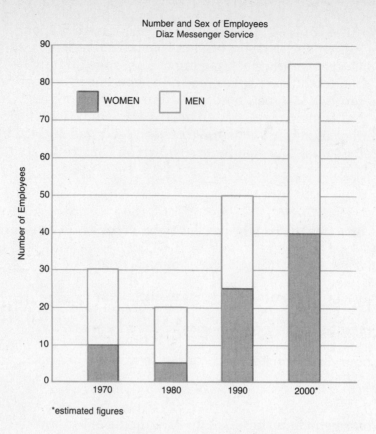

Number and Sex of Employees
Diaz Messenger Service

Figure 3.5  An Extended Bar Graph

The key tells us the shaded part of the bar stands for women employees.

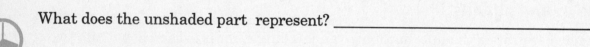

What does the unshaded part  represent? _____

_____

The unshaded part of the bar represents men. The entire bar, shaded and unshaded, represents all the employees of Diaz Messenger Service, both women and men.

## READING AN EXTENDED BAR GRAPH

You read an extended bar graph much the same way you read any other bar graph. For example, suppose you want to find out how many people in all were employed by the company in 1970.

First, find the bar representing that year. Then trace across the graph from the end of that bar to the vertical scale labelled Number of Employees.

How many people worked at Diaz in 1970? _____

According to the graph, 30 people were employed there in 1970.

Now suppose you want to know how many of those 30 employees were women. To find this number, work only with the shaded part of the bar. Find the top of the shaded part of the bar and trace across the graph to the scale on the left.

Of the 30 employees in 1970, how many were women? _____

The graph shows that 10 of the 30 employees were women.

To learn how many men were employed at Diaz in 1970 will require one more step.

First, find the total number of employees for that year. It is 30. Next, subtract the number of women workers, which is 10. The difference is the number of male employees: 30 − 10 = 20.

## WARMUP C

**Use the extended bar graph in Figure 3.5 to answer these questions.**

1. How many people in all were employed at Diaz in 1980? _____

2. How many women worked at Diaz in 1980? (Because the bar falls between two

   measures on the left scale, you will have to estimate the number.) _____

3. How many people in all were employed at Diaz in 1990? _____

4. During one year, the same number of women as men worked at Diaz. What was that

   year? _____

5. How many more workers does Diaz estimate it will have in the year 2000 than it had

   in 1990? _____

**Check your answers on page 148.**

## MULTIPLE BAR GRAPHS

Some bar graphs use more than one bar in the same scale. For example, the same information presented in the extended bar graph in Figure 3.5 (see page 26) could also be presented in a **multiple bar graph**. It would then look something like Figure 3.6.

Now the difference in the number of men and women working at Diaz in a given year becomes even clearer. What is more, the year in which the number of women employees was the same as the number of men—1990—is easily determined since both bars are the same height. However, finding the total number of employees in any year takes an extra step. Now you must first find the number of men, then find the number of women, and then add the two to get the answer.

Figure 3.7 is an example of a multiple bar graph.

 What kind of information does the graph show? _____

_____

The graph shows changes in the population of the United States from 1900 to 1980. For each time period there are two bars—one representing the number of people living in rural (farm) areas, and the other showing the number of people living in urban (city) areas.

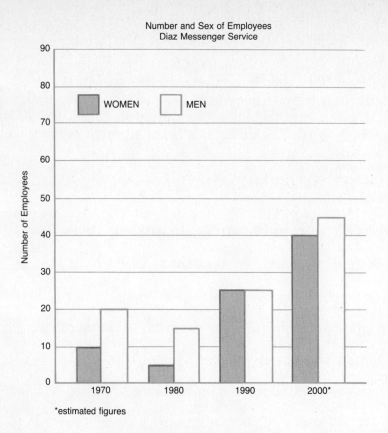

Figure 3.6  A Multiple Bar Graph

 What does the vertical scale measure? _____

_____

The vertical scale measures the population in millions of people. That is, 40 on the scale represents 40 million people.

 What does the horizontal scale measure? _____

_____

The horizontal scale tells us in which year the population was counted.

## READING A MULTIPLE BAR GRAPH

By measuring the height of a bar, you can tell how many people lived in rural and urban areas. For example, look at the two bars in the scale labeled 1900. To find out how many people lived in rural areas of the United States in that year, trace across from the top of the bar representing rural population to the vertical scale. Because the top of the bar lies between 40 and 60 million on the scale, you will have to estimate.

It appears that rural Americans numbered about 43 million in 1900. The top of the bar lies much closer to 40 million than to 60 million on the scale. The number 43 million appears to be a good estimate.

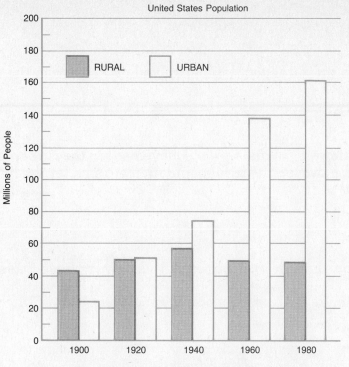

Figure 3.7  Example of a Multiple Bar Graph

About how many people lived in urban areas in 1900? _____

You could estimate the number to be about 24 million people.

How would you find the *total* population, both rural and urban, for 1900?

_____

To find the total population for 1900, add the rural population (43 million) to the urban population (24 million): 43 + 24 = 67. In 1900 there were 67 million people living in the United States.

## DISCOVERING TRENDS ON A MULTIPLE BAR GRAPH

Trends, you may recall from page 21, are upward or downward movements over a period of time. For example, look at the bars on the graph in Figure 3.7.

Do the bars reveal an upward or downward trend in the urban population of the United States? (Remember, the longer the bar, the greater its value.)

_____

_____

According to the graph the urban population of the United States showed a strong upward trend during the years 1900 to 1980.

 What trend in the rural population can you discover from the same graph?
_____

Here the graph shows a gradual upward movement in the rural population between 1900 and 1940. After that, however, the rural population got smaller as shown by the shorter bars in 1960 and in 1980.

## WARMUP D

**Use the multiple bar graph in Figure 3.7 to answer these questions.**

1. What is your estimate of the rural population of the United States in 1920?

   _____

2. What is your estimate of the urban population in 1920? _____

3. Based on those estimates, what is your estimate of the total population in 1920?

   _____

4. What is your estimate of the total population of the United States in 1960?

   _____

5. In 1980, about how many more people lived in urban areas than lived in rural areas?

   _____

**Check your answers on page 148.**

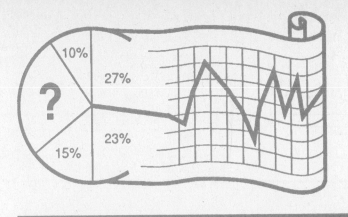

## How Carefully Did You Read?

**A. Choose the correct completion for each statement.**

1. A bar graph always has

   ☐ a. two scales.
   ☐ b. pictures.
   ☐ c. a key.

2. Bar graphs usually

   ☐ a. show temperatures.
   ☐ b. use symbols.
   ☐ c. compare sets of figures.

3. A graph on which the bars run up and down is known as a

   ☐ a. vertical bar graph.
   ☐ b. horizontal bar graph.
   ☐ c. chart.

4. On a bar graph, the shorter the bar, the

   ☐ a. greater its value.
   ☐ b. less accurate the graph is.
   ☐ c. lower its value.

5. The purpose of a bar graph is always found in the

 ☐ a. title.

 ☐ b. source.

 ☐ c. key.

6. An extended bar graph

 ☐ a. has longer bars.

 ☐ b. is the same as a multiple bar graph.

 ☐ c. has more than one segment within each bar.

7. Trends are

 ☐ a. found only in multiple bar graphs.

 ☐ b. ways of telling how accurate a bar graph is.

 ☐ c. upward or downward movements over a period of time.

8. On a vertical bar graph showing an upward trend, the bars would

 ☐ a. grow generally shorter across the graph.

 ☐ b. grow generally longer across the graph.

 ☐ c. remain the same length across the graph.

**B. Use the graph in Figure 3.8 to complete the statements.**

1. Figure 3.8 is an example of a _____ bar graph.

2. The graph shows _____

   _____.

3. The cost of a 30-second ad on Channel 10 in 1989 was _____.

4. The cost of a 30-second ad on Channel 12 in 1990 was _____.

5. The cost of a 30-second ad on Channel 6 in 1991 was _____.

6. In 1989 a 30-second ad on all three television stations during the 6:00 P.M. news cost

   _____.

7. Thirty-second ads on all three stations in 1991 cost _____.

8. According to the graph, the cost of a 30-second ad on Channel 6 shows a/an

   _____ trend over the three-year period.

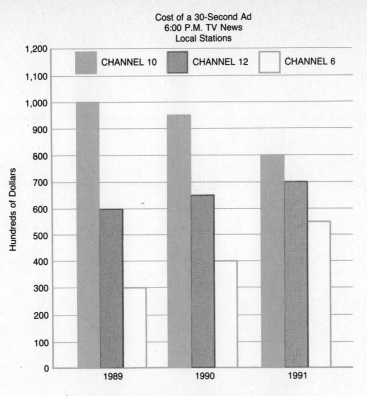

Figure 3.8   Cost of a 30-Second TV Ad

9. In 1990, the cost of a 30-second ad on Channel 10 was _____ dollars more
   than on Channel 12.

10. According to the graph, the cost of a 30-second ad on Channel 10 shows a/an
    _____ trend over the three-year period.

**Check your answers on page 148.**

> **Look** back at How Much Do You Already Know? on page 20.
> Did you complete each statement correctly? If not, can you do so
> now?

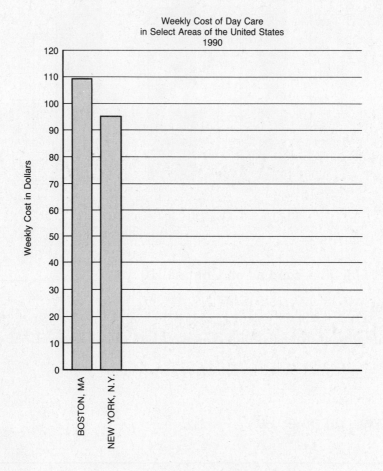

Weekly Cost of Day Care
in Select Areas of the United States
1990

LOCATION OF DAY CARE CENTERS

Figure 3.9  Completing a Bar Graph

ARE YOU READY FOR THE CHALLENGE of making a bar graph from information given you? If you think you are, use the data that follow to construct a vertical bar graph.

In constructing the graph in Figure 3.9, you will have to estimate the length of most of the bars. Be sure the top of your bar falls between the numbers on the scale above and below the actual figure you wish to represent. The first two have been done for you.

Day care for children remains a growing need in the United States. The 1990 figures for the ten metropolitan areas with the highest weekly day care costs were as follows: Boston, Massachusetts: $109; New York, New York: $95; Anchorage, Alaska: $91; Manchester, New Hampshire: $90; Washington, District of Columbia: $87; Minneapolis, Minnesota: $87; Hartford, Connecticut: $86; Philadelphia, Pennsylvania: $86; Portland, Maine: $83; Burlington, Vermont: $79.

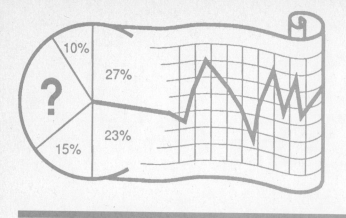

## How Much Do You Already Know?

**Choose the correct completion for each statement. If you are not sure about an answer, do not guess.**

1. A line graph will always have

   ☐ a. two lines.

   ☐ b. two scales.

   ☐ c. one key.

2. The first step in reading a line graph is

   ☐ a. to check the source.

   ☐ b. to trace across the graph to the vertical scale.

   ☐ c. to find out what the graph shows.

3. The horizontal scale on a line graph almost always

   ☐ a. measures a unit of time.

   ☐ b. tells how accurate the graph is.

   ☐ c. is more important than the vertical scale.

4. A multiple line graph

   ☐ a. is more accurate than a single line graph.

   ☐ b. has more scales than a single line graph.

   ☐ c. allows us to compare two or more sets of numbers.

**Check your answers on page 148.**

# Line Graphs

Like bar graphs, **line graphs** show trends in a way that makes the trends easy to understand. However, the use of a line rather than a set of bars means that the information on a line graph is presented in a continuous manner. The information on a line graph is not broken up into disconnected segments of data as on a bar graph.

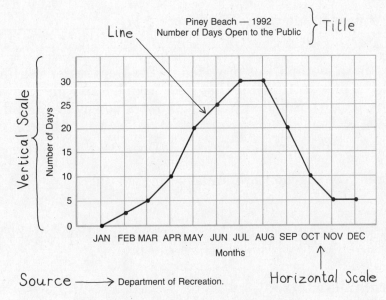

**Figure 4.1  The Parts of a Line Graph**

Look at Figure 4.1. What does this line graph show? _____

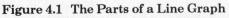

The graph shows the number of days Piney Beach was open to the public during 1992.

## PARTS OF A LINE GRAPH

Like tables, charts, and bar graphs, line graphs have titles, source lines, and scales. Like bar graphs, line graphs always have two scales: A horizontal scale and a vertical scale.

The vertical scale can measure many things—dollars, temperatures, or percentages, for example.

 What does the vertical scale in Figure 4.1 measure? _____

_____

This scale provides a measure of the number of days the beach was open each month. (The scale is divided into five-day increments, or steps.)

The horizontal scale runs across the bottom of the graph. Horizontal scales almost always measure units of time—hours, days, weeks, months, or years, for example.

 What does the horizontal scale in Figure 4.1 measure? _____

_____

This horizontal scale shows months, from January through December.

Many line graphs will include a source. The source tells where the information comes from. Look for the source at the bottom of the graph.

What is the source of information in Figure 4.1?_____

The information in this graph was provided by the Department of Recreation.

Finally, there is the line itself. A line that rises from left to right shows an upward trend. A line that descends, or falls, from left to right shows a downward trend.

 Look at the line on the graph in Figure 4.1. Does the line show an upward or a

downward trend for the months January through July? _____

Since the line rises for those months, it indicates a rising trend in the number of days the beach was open.

## READING A LINE GRAPH

The first step in reading a line graph is to find out what the graph shows. The title will tell you this. Next look at the two scales to see what units of measure the graph uses. Read the vertical scale from the bottom up. Read the horizontal scale from left to right.

Next, quickly glance at the line. The line is always read from left to right. Does it show a rising or a falling trend overall? Or does it show a rising and then a falling trend?

Finally, see if the graph has a source. As in any kind of graph, the source in a line graph will give you an idea of how reliable and accurate the figures are.

Now suppose you want to know how many days Piney Beach was open during the month of April. Follow these steps:

- Look across the horizontal scale to find the month labeled April.
- Follow the line for April straight up until you reach the mark on the line graph for that month.

- Trace across to the vertical scale. The number you read tells how many days the beach was open in April.

 How many days was the beach open in April? _____

If you followed the steps accurately, you learned that the beach was open 10 days in April.


## WARMUP A

**Use the line graph in Figure 4.1 to answer these questions.**

1. How many days was the beach open in March? _____

2. How many days was the beach open in July? _____

3. The beach was open 20 days in two separate months. What were those months?

   _____ and _____

4. How many *more* days was the beach open in September than in November?

   _____

5. Does the line representing the number of days the beach was open between August

   and December show an upward or a downward trend? _____

**Check your answers on page 148.**


## ESTIMATING AMOUNTS ON A LINE GRAPH

In reading a line graph, you will notice that some points on the line may fall between two units of measure on the vertical scale. For example, look at the highest point in the line graph in Figure 4.2.

According to the title, this graph shows the number of classified advertisements placed each month in the *Daily Herald-Tribune* newspaper.

 What is the source of the information? _____

The source is the Classified Advertising Department of the newspaper.

 Notice the two scales on the graph. What does the vertical scale measure?

_____

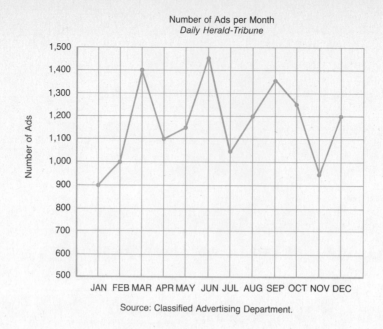

Number of Ads per Month
*Daily Herald-Tribune*

Source: Classified Advertising Department.

**Figure 4.2  A Single Line Graph**

The vertical scale, to the left of the graph, measures the number of ads. The scale runs from 500 ads to 1500 ads in increments of one hundred.

 What does the horizontal scale measure? _____

The horizontal scale, which runs from left to right across the bottom of the graph, shows the months in which the ads were placed.

 How many ads were placed in the newspaper in January? _____

Nine hundred ads were placed in January.

To arrive at that answer, you first had to find January on the horizontal scale. Then you followed the line for that month straight up until you reached the mark on the line for that month. Finally, you traced across the vertical scale on the left, to read the number 900.

Now suppose you want to find out how many ads were placed in May. If you follow the steps in reading a line graph, you will discover that the line representing May falls somewhere between two units of measure on the vertical scale.

 What are those two units of measure? _____ and _____

The mark representing May on the line falls between 1,100 and 1,200 ads. This is why you will need to estimate. Looking more closely at the mark, you will see it falls about halfway between 1,100 and 1,200 on the vertical scale. A good estimate for the number of ads placed in May is 1,150. (The difference between 1,100 and 1,200 is 100. Half of this is 50, which is added to the lower figure: (1,100 + 50 = 1,150.)

Use the line graph in Figure 4.2 to answer these questions.

1. How many ads were placed in the newspaper in February? _____

2. How many *more* ads were placed in February than were placed in January?

   _____

3. Which three months saw the most ads placed? _____

4. How many ads were lost between October and November? _____

5. In all, how many ads were placed in the newspaper during July and August?

   _____

**Check your answers on page 148.**

## MULTIPLE LINE GRAPHS

Some line graphs have two, or even more, lines. These are called **multiple line graphs**. Multiple line graphs allow information from two or more sets of figures to be compared.

Look at the multiple line graph in Figure 4.3. It contains all the information on the line graph in Figure 4.2. In addition, there is a second line on this graph—a green line.

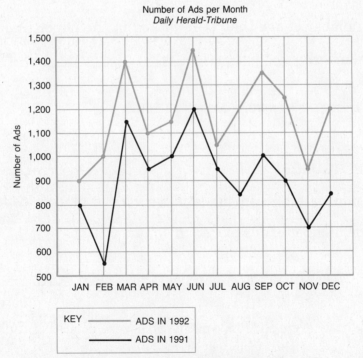

Figure 4.3  A Multiple Line Graph

What does the green line represent? (Look for the key at the bottom of the graph). _____

The key tells us the green line represents ads placed in 1991. (Notice that the key also tells us that the black line represents ads placed in 1992.)

Always look for a key in multiple line graphs. The key will explain what the various lines in the graph represent.

## READING A MULTIPLE LINE GRAPH

Read a multiple line graph the same way you read a single line graph. Just be sure you use the correct line in determining the values. For example, how many ads were placed in the *Daily Herald-Tribune* in January 1991?

- Look across the horizontal scale for the month labeled January.
- Follow the line for January straight up until you reach the mark on the *green* line for that month in 1991.
- Trace across to the vertical scale. The number you read tells you how many ads were placed in January 1991.

 How many ads were placed in the paper in January 1991? _____

According to the graph, 800 ads were placed in January 1991.

 Now find the number of ads that were placed in the paper in January 1992. What is that number? _____

The graph shows that 900 ads were placed in 1992.

## COMPARING SETS OF FIGURES

 Was there a gain or a loss in the number of ads placed in January 1992 compared to the number of ads placed in January 1991? _____

The figures show a gain of 100 ads. To arrive at that answer, subtract (900 − 800 = 100).

Find February on the graph. How many ads were placed in February 1991? (You will have to estimate, since the green line falls between two units of measure on the vertical scale.) _____

The graph shows that 550 ads were placed in February 1991.

Now find the number of ads placed in February 1992. _____

Here the graph shows 1,000 ads were placed in February 1992.

What was the *increase* in the number of ads placed in February 1992 compared to the number placed in February 1991? _____

Four hundred and fifty more ads were placed in February 1992 than were placed in February 1991 (1,000 − 550 = 450).

**WARMUP C**

**Use the multiple line graph in Figure 4.3 to answer these questions.**

1. How many ads were placed in the *Daily Herald-Tribune* in September 1991?

   _____

2. How many ads were placed in September 1992? _____

3. How many *more* ads were placed in September 1992 than in September 1991?

   _____

4. Which month in 1991 showed the largest *increase,* or gain in ads placed? (Look for the longest rising green line from one month to the next.) _____

5. Overall, did the *Daily Herald-Tribune* run more or fewer ads in 1992 than in 1991?

   _____

**Check your answers on page 148.**

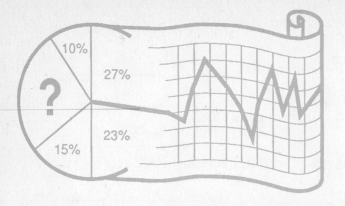

## How Carefully Did You Read?

**A. Choose the correct completion for each statement.**

1. The vertical scale on a line graph is found

    ☐ a. at the left side of the graph.
    ☐ b. across the bottom of the graph.
    ☐ c. next to the key.

2. One difference between a single line graph and a multiple line graph is that the

    ☐ a. single line graph has no vertical scale.
    ☐ b. multiple line graph always has a key.
    ☐ c. single line graph sometimes lacks a source.

3. On a line graph, horizontal scales almost always measure

    ☐ a. units of time.
    ☐ b. the length of a line.
    ☐ c. amounts of money.

4. The lines in line graphs are always read from

    ☐ a. top to bottom.
    ☐ b. left to right.
    ☐ c. bottom to top.

5. The part of a multiple line graph that tells you the difference between the two lines is the

☐ a. title.
☐ b. source.
☐ c. key.

6. On a line graph, a line that goes steadily up shows a

☐ a. falling trend.
☐ b. flat, or even, trend.
☐ c. rising trend.

7. In reading a line graph, if the line falls between two units of measure on the vertical scale, you must

☐ a. use the lower number as your answer.
☐ b. use the higher number as your answer.
☐ c. estimate a number between the two as your answer.

8. Line graphs that have two or more lines are called

☐ a. single line graphs.
☐ b. extended line graphs.
☐ c. multiple line graphs.

**B. Use the graph in Figure 4.4. to complete the statements.**

1. Figure 4.4 is an example of a _____ graph.

2. The graph shows _____.

3. The graph shows an income of _____ for the Two Left Feet Shoe

   Company in 1989.

4. In 1989, expenses amounted to _____.

5. The greatest income was earned in _____.

6. The lowest income was earned in _____.

7. Expenses were greater than income in the years _____and _____.

8. The total income for the shoe company for all five years shown on the graph amounts

   to_____.

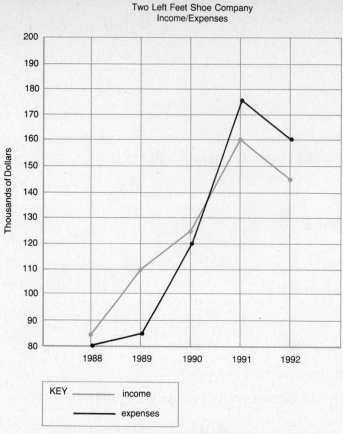

Two Left Feet Shoe Company
Income/Expenses

Thousands of Dollars

1988  1989  1990  1991  1992

KEY ——— income
    ——— expenses

Source: Tally Accountants.

**Figure 4.4  Income and Expenses Graph**

9. The total expenses for the five years amount to _____.

10. If expenses continue to exceed income in the years 1993 and beyond, the Two Left

    Feet Shoe Company is likely to _____.

**Check your answers on page 148.**

Look back at How Much Do You Already Know? on page 36. Did you complete each statement correctly? If not, can you do so now?

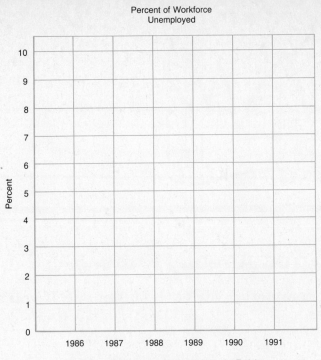

Percent of Workforce
Unemployed

Source: Department of Employment and Training

**Figure 4.5  Filling in the Parts of a Line Graph**

ARE YOU READY FOR THE CHALLENGE of making a line graph from information given you? If you think you are, use the figures that follow to construct a single line graph.

Plot the figures on the graph in Figure 4.5. Then connect the marks with a solid line to complete the graph.

Unemployment showed a rising trend after several years of steady decline. In 1986, 4% of the state's workforce was unemployed. This declined to 3½% in 1987 and to 3% in 1988. By 1989 only 2½% of the workforce was unemployed. This figure rose steeply in 1990, however, to 7%, then to 7½% in 1991.

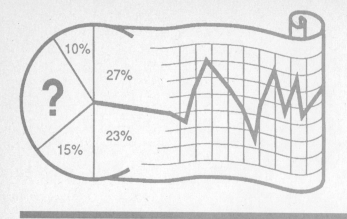

## How Much Do You Already Know?

Choose the correct completion for each statement. If you are not sure about an answer, do not guess.

1. Combined bar and line graphs

   ☐ a. replace simple tables.

   ☐ b. make use of row headings and column headings.

   ☐ c. compare two sets of figures.

2. One common kind of combined bar and line graph shows

   ☐ a. mileages between several cities.

   ☐ b. temperature and precipitation in a certain area.

   ☐ c. populations of states and cities.

3. Unlike other graphs, a combined bar and line graph will have

   ☐ a. one scale.

   ☐ b. two scales.

   ☐ c. three scales.

4. On a combined bar and line graph, the line always runs

   ☐ a. up and down.

   ☐ b. from left to right.

   ☐ c. in a straight line.

Check your answers on page 149.

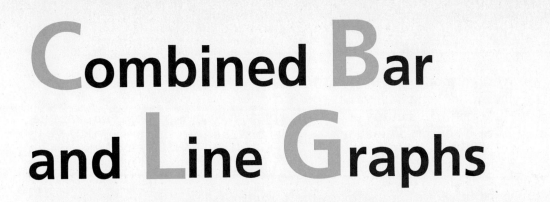

# Combined Bar and Line Graphs

## PARTS OF A COMBINED BAR AND LINE GRAPH

Multiple line graphs (Chapter 4) offer one way to compare two sets of figures. Graphs which include both a bar graph and a line graph offer another.

In a **combined bar and line graph**, the bars run vertically, or up and down. The line always runs from left to right.

One common kind of combined bar and line graph shows temperatures and precipitation (rain, snow, sleet) amounts over a period of time. Look at Figure 5.1.

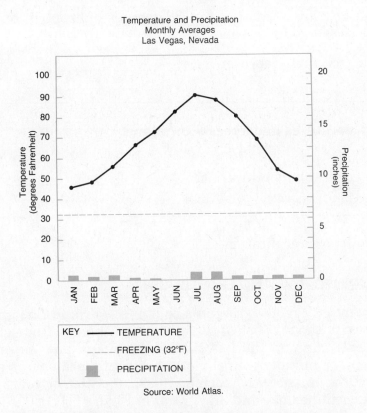

Figure 5.1  The Parts of a Combined Bar and Line Graph

As in any graph, the kind of information that a combined bar and line graph shows can be found in the title.

What kind of information does the graph in Figure 5.1 show? _____

_____

Figure 5.1 shows average monthly temperatures and amounts of precipitation for a complete year for the city of Las Vegas, Nevada.

A combined bar and line graph has many of the same parts as other graphs. For example, besides a title, it may have a source. The source tells where the information in the graph comes from.

 What is the source of information in the graph in Figure 5.1? _____

The source, found at the bottom of the graph, tells us the information was taken from a world atlas.

A combined bar and line graph also has a key. The key explains any symbols used on the graph.

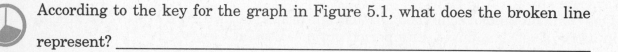

 According to the key for the graph in Figure 5.1, what does the broken line

represent? _____

It represents the freezing point of water (32 degrees Fahrenheit or 0 degrees Celsius).

 What does the solid line represent? _____

_____

It represents temperatures in degrees Fahrenheit.

 What do the vertical bars represent? _____

_____

They show amounts of precipitation.

## SCALES ON A COMBINED BAR AND LINE GRAPH

The most important difference between a combined bar and line graph and other kinds of graphs has to do with the number of scales. Line graphs and bar graphs each have two scales. A vertical scale runs up the left side of the graph and measures such things as temperatures, dollars, or numbers of people. A horizontal scale runs across the bottom. It almost always measures units of time such as weeks, months, or years. Both of these scales can be found on the combined bar and line graph in Figure 5.1.

Look at the vertical scale on the graph. What does it measure? _____

_____

The vertical scale measures temperatures in degrees Fahrenheit, from 0 to 100 in 10-degree increments.

 What does the horizontal scale on the same graph measure? _____

_____

The horizontal scale shows the months from January through December.

Now look to the right of the graph in Figure 5.1. There you will see a third scale.

 What does this third scale measure? _____

_____

This vertical scale measures precipitation in inches. Each **tick mark** (—) stands for one inch of precipitation. To avoid cluttering the scale with numbers, only 5, 10, 15, and 20 (inches) appear on the scale.

## READING A COMBINED BAR AND LINE GRAPH

Reading a combined bar and line graph is no harder than reading any bar or line graph. But you must be sure to choose the scale that applies to the part of the graph you are reading.

For example, when reading a temperature line, use the vertical scale that measures temperatures. In the graph in Figure 5.1, this is the scale at the *left* of the graph.

Similarly, in reading the graph to find the amount of precipitation, use the vertical scale that measures precipitation. In Figure 5.1 this is the scale at the *right* of the graph.

Now suppose you are planning a trip to Las Vegas in April. How hot will it be? How much precipitation might you expect that month? First, look at the temperature.

- Trace across the horizontal scale at the bottom of the graph until you find April.
- Follow straight up until you reach the mark on the temperature line for that month.
- Trace across to the vertical scale at the *left* of the graph. The number you read tells you the average April temperature.

 What is the average temperature for April in Las Vegas? (You must estimate to

find the answer.)_____

According to the graph, the average temperature is 65°. (The temperature lines falls about halfway between 60° and 70° on the vertical scale, so 65° is a good estimate.)

To find the average amount of precipitation for April, follow these steps.

- Trace across the horizontal scale to find April.
- Find the top of the bar for April.

- Now trace across to the vertical scale at the *right* of the graph. The figure you read tells you the amount of precipitation you can expect in April.

 What is the average precipitation in Las Vegas for April? (You must estimate to find the answer.) _____

According to the graph, you can expect about one-quarter of an inch of precipitation that month. No need to take a raincoat or umbrella on your trip!

**WARMUP**

**Use the graph in Figure 5.1 to answer these questions.**

1. What is the average temperature in Las Vegas in July? _____

2. How much precipitation might you expect in July? _____

3. In what month does the lowest average temperature occur? _____

4. Assuming that you like hot weather and as little chance of precipitation as possible,

   which month would you choose to visit Las Vegas? _____

5. What is the difference in average temperature between the warmest and coolest

   months in Las Vegas? _____

**Check your answers on page 149.**

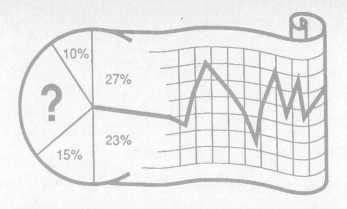

## How Carefully Did You Read?

**A. Choose the correct completion for each statement.**

1. Combined bar and line graphs offer a way to

   ☐ a. check the accuracy of other kinds of graphs.

   ☐ b. compare two sets of figures.

   ☐ c. simplify the information shown on a bar or a line graph.

2. Combined bar and line graphs have

   ☐ a. none of the features of other kinds of graphs.

   ☐ b. many of the same features as other graphs.

   ☐ c. many features not found on other graphs.

3. Combined bar and line graphs have

   ☐ a. one scale.

   ☐ b. two scales.

   ☐ c. three scales.

4. The key on a combined bar and line graph

   ☐ a. tells how accurate the graph is.

   ☐ b. tells where the information on the graph came from.

   ☐ c. explains any symbols used on the graph.

5. On a combined bar and line graph, the two vertical scales always measure

☐ a. the same thing.
☐ b. in increments of ten.
☐ c. two different things.

6. On a combined bar and line graph, the horizontal scale almost always measures

☐ a. units of time.
☐ b. amounts of precipitation.
☐ c. population growth or decline.

**B. Use the graph in Figure 5.2 to complete the statements.**

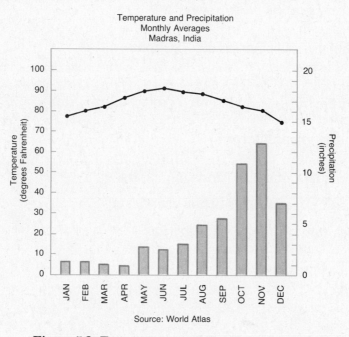

Figure 5.2  Temperature and Precipitation Graph

1. During the hottest month in Madras, India, the temperature is likely to reach at least _____ degrees.

2. During the coolest month, the average temperature is _____ degrees.

3. The average temperature in Madras in October is _____ degrees.

4. The month with the least amount of precipitation is _____.

5. The month that has the most precipitation is _____.

6. The difference in amounts of precipitation between the wettest and driest months is about _____ inches. (You must make sure the fractions have the same denominator before you subtract to find the answer.)

7. Which month has an average temperature of 80° and 1 inch of precipitation?

_____

8. The total average precipitation for Madras is about _____ inches a year. (Before you add to find the answer, be certain all fractions have the same denominator.)

9. In October about _____ more inches of precipitation fall than in July.

10. You would not expect to find any ski resorts in Madras, India, because _____

_____.

**Check your answers on page 149.**

Look back at How Much Do You Already Know? on page 48. Did you complete each statement correctly? If not, can you do so now?

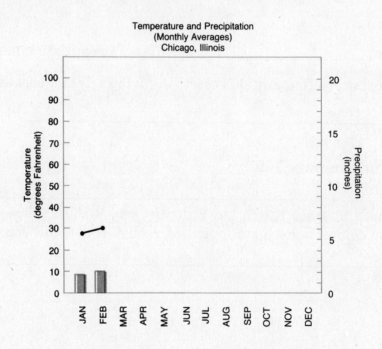

**Figure 5.3  Completing a Combined Bar and Line Graph**

ARE YOU READY FOR THE CHALLENGE of making a combined bar and line graph from information given you? If you think you are, use the figures that follow to complete the graph in Figure 5.3. The first two months have been done for you.

Here are the average temperatures for the city of Chicago, Illinois.* January, 28°; February, 30°; March, 38°; April 50°; May, 60°; June, 70°, July, 75°; August, 79°; September, 68°; October, 58°; November, 40°; December, 30°.

The average precipitation for a year is as follows. January, 1¾ inches; February, 2 inches; March, 3 inches; April, 3½ inches; May, 3¾ inches; June, 4 inches; July, 3½ inches; August, 3 inches; September, 2¼ inches; October, 3 inches; November, 2 inches; December, 2 inches.

*Source: World Atlas.

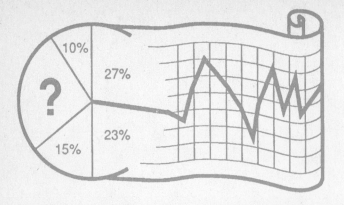

## How Much Do You Already Know?

**Choose the correct completion for each statement. If you are not sure about an answer, do not guess.**

1. Circle graphs, always

   ☐ a. use symbols.

   ☐ b. show trends.

   ☐ c. show parts of a whole.

2. Circle graphs are also known as

   ☐ a. pie charts.

   ☐ b. bar graphs.

   ☐ c. line graphs.

3. On a circle graph that pictures sections as percents, all the sections will add up to

   ☐ a. 100%.

   ☐ b. 75%.

   ☐ c. 50%.

4. On a circle graph that shows that a person spends 5% of a $1,000 budget on telephone calls, the 5% represents

   ☐ a. $1,000.

   ☐ b. $500.

   ☐ c. $50.

**Check your answers on page 149.**

# Circle Graphs

## Parts of a Circle Graph

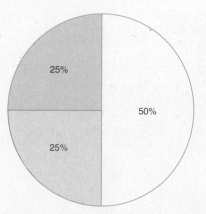 We have seen that tables and charts present their information in columns and rows. They provide us with an efficient way to compare many individual bits of information. Bar graphs and line graphs are useful in showing trends—upward and downward movements over a period of time.

**Circle graphs** give a clear picture of the parts of a whole and how each part compares to the others.

Sometimes each part of a circle graph is used to represent a percent of the whole. In this case each segment of the graph is labeled with a percentage. The parts will always add up to 100%, as shown in Figure 6.1.

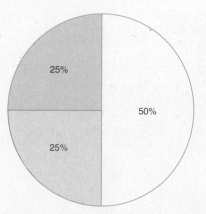

**Figure 6.1  The Parts of a Circle Graph**

Sometimes the parts of a circle graph are used to break down a large number into several smaller numbers. In this case, the total of the smaller numbers will always add up to the larger number, as shown in Figure 6.2.

Each section of a circle graph looks like a slice of a pie, which is why circle graphs are sometimes known as **pie charts**.

### CIRCLE GRAPHS THAT SHOW PERCENTS

Look at the circle graph in Figure 6.3.

A circle graph always has a title, and often a source. As with other kinds of graphs, the title tells what the graph shows. Look for the title above the graph.

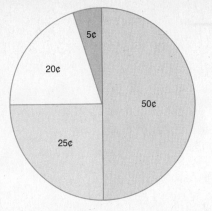

**Figure 6.2 "Slices" of a "Pie" Graph**

What is the title of the graph in Figure 6.3? _____

_____

The title is Who Are the Homeless?

On a circle graph the source tells where the information on the graph came from. Look for the source at the bottom of the graph.

What is the source of information for the graph in Figure 6.3? _____

_____

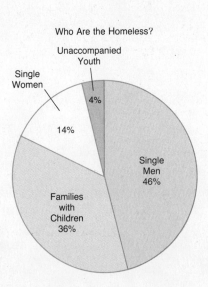

Source: United States Conference of Mayors.

**Figure 6.3 Circle Graph Using Percents**

The source is the United States Conference of Mayors.

 Notice that the graph in Figure 6.3 has four sections, or slices. What do these

sections represent? _____  _____  _____

_____

Each section in the graph represents one segment of the homeless population in the United States: Single Men, Families with Children, Single Women, and Unaccompanied Youth.

## READING A CIRCLE GRAPH

In reading any circle graph first look for the title to learn what the graph shows. Next, look to see if the graph has a source. The source suggests how accurate and reliable the information on the graph is. Then look at the graph itself. The size of each section on the graph tells you something about the relative size of that part in relation to the the whole thing. Notice that the sections show the percent of people, not the number of people, in each category.

 What percent of the homeless population are single women? _____

The graph shows that 14% of the homeless population is made up of single women.

## WARMUP A

**Use the graph in Figure 6.3 to answer these questions.**

1. What percent of the homeless population is made up of families with children?

   _____

2. What percent of the homeless population are single men? _____

3. What group accounts for the smallest percentage of the homeless? _____

4. What is the difference between the percentage of unaccompanied youth and the

   percentage of single women? _____

5. What is the total percentage of homeless shown on the graph? _____

**Check your answers on page 149.**

## CIRCLE GRAPHS THAT SHOW AMOUNTS

The sections of the graph in Figure 6.4 represent parts of a dollar.

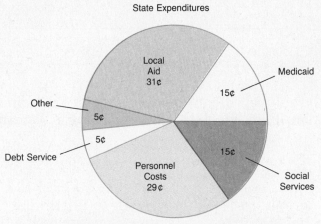

Source: Massachussetts State Budget Bureau, 1989.

**Figure 6.4  Circle Graph Showing Amounts**

What is the title of the graph in Figure 6.4? _____

The title is State Expenditures.

The graph shows how the state spends each dollar in its budget. It does not tell how many dollars in all were spent.

What part of each dollar did the state spend on Medicaid? _____

According to the graph, 15¢ of each dollar was spent on Medicaid.

## WARMUP B

**Use the circle graph in Figure 6.4 to answer these questions.**

1. How much of each dollar did the state spend on personnel costs (salaries and benefits

   for its workers)? _____

2. What segment of expenditures took the largest amount of each dollar? _____

3. What part of each dollar spent was used for Medicaid and Social Services together?

   _____

4. What part of each dollar did the state spend on aid to cities and towns? _____

**5.** How much do the amounts shown in all the sections add up to? _____

**Check your answers on page 149.**

## INTERPRETING CIRCLE GRAPHS

Look at the circle graph in Figure 6.5.

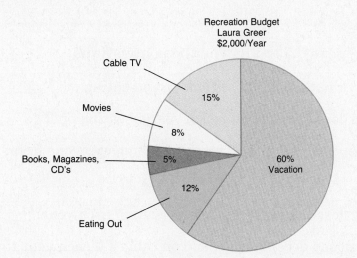

**Figure 6.5   Circle Graph of Percents and Amounts**

What does this graph show? _____

_____

As the title indicates, the graph shows how Laura Greer spends her annual recreation budget of $2,000.

What percent of her budget does she spend on movies?_____

The graph shows she spends 8% of her annual recreation budget on movies.
Suppose you want to know how much money this 8% represents. The title of the graph tells us the total amount of her recreation budget is $2,000. The section entitled Movies tells us she spent 8% of this $2,000 on movies. To find out how much 8% of $2,000 is, you must multiply. To do this

Change the percent to its decimal equivalent: 8% = .08
Multiply .08 times $2,000: 2,000 × .08 = 160

The result, $160, is the amount of money she spends on movies each year.

What percent of her annual recreation budget does Laura spend on books,

magazines, and CDs?_____

The graph shows she spends 5% of her budget on those items.

 In dollars, how much of her budget does the 5% represent? _____

If you multiply the decimal equivalent of 5% (.05) times $2,000 (her total annual recreation budget), you will see she spends $100 on books, magazines, and CDs each year.

**WARMUP C**

**Use the graph in Figure 6.5 to answer these questions.**

1. What percent of her budget does Laura spend on eating out? _____

2. How much money does Laura spend each year on eating out? _____

3. Each year Laura spends 60% of her recreation budget on one item. What is that item?

   _____

4. How much money does Laura budget for a vacation each year? _____

5. How much more money does Laura spend on cable TV each year than she spends on

   books, magazines, and CDs? (You must first multiply to find amounts of money, then

   subtract to find the answer.) _____

**Check your answers on page 149.**

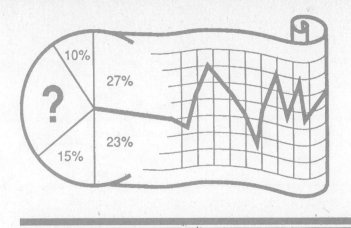

## How Carefully Did You Read?

**A. Choose the correct completion for each statement.**

1. Circle graphs are useful in

   ☐ a. showing changes over a period of time.
   ☐ b. comparing parts of a whole.
   ☐ c. showing continuous data.

2. The source on a circle graph can be important because it shows

   ☐ a. who drew the graph.
   ☐ b. where the information on the graph came from.
   ☐ c. the subject of the graph.

3. In reading a circle graph, you should first look at the

   ☐ a. title.
   ☐ b. sections of the graph.
   ☐ c. source.

4. When each part of a circle graph is used to show a percent of a whole, each segment will contain

   ☐ a. a dollar figure.
   ☐ b. a percentage.
   ☐ c. more than one number.

5. On a circle graph a 20% section is

☐ a. smaller than a 12% section.

☐ b. about the same size as a 12% section.

☐ c. larger than a 12% section.

6. When a circle graph shows that a person spends 12% of a $4,000 budget on automobile expenses, the 12% represents

☐ a. $480.

☐ b. $4,800.

☐ c. $48,000.

## B. Use the graphs in Figure 6.6 to complete the sentences.

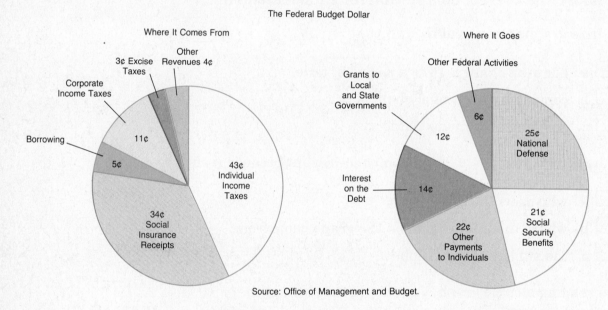

Figure 6.6 A Pair of Circle Graphs

1. The overall title of the two graphs is _____

_____.

2. The graph on the left shows _____.

3. The graph on the right shows _____.

4. The graph on the left shows that _____ of each tax dollar comes from

corporate income taxes.

5. According to the graph on the left, the greatest part of every budget dollar comes from

_____.

6. According to the graph on the left, the smallest part of every budget dollar comes

   from _____.

7. The graph on the left shows that after all taxes and receipts have been collected, the

   government still has to borrow _____ to make up the budget dollar.

8. The graph on the left shows that the government collects _____ more of

   each tax dollar from individual income taxes than it does from corporate income taxes.

9. According to the graph on the right, _____ of every tax dollar goes to

   paying off interest on the national debt.

10. For every dollar the government takes in, it spends _____ on national

    defense.

11. Social security benefits and other payments to individuals make up _____

    of every federal budget dollar spent.

12. The graph on the right shows that the government spends _____ less on

    grants to local and state governments than it does on interest on the national debt.

13. In all, _____ of every budget dollar goes to national defense and other

    federal activities.

14. Of the federal budget dollar, 6¢ is spent on other federal activities. The 6¢ represents

    _____ percent of the federal dollar.

**Check your answers on page 149.**

Look back at **How Much Do You Already Know?** on page 58.
Did you complete each statement correctly? If not, can you do so
now?

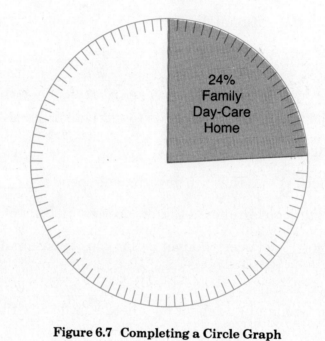

24%
Family
Day-Care
Home

**Figure 6.7 Completing a Circle Graph**

**ARE YOU READY FOR THE CHALLENGE** of making a circle graph from information given you? If you think you are, use the data that follow to fill in the circle graph in Figure 6.7. Be sure to give the graph a title and a source.

Every circle graph is divided into 100 equal sections. In Figure 6.7 each section equals 1% of the whole circle.

According to the Office of Management and Budget of the federal government, 47% of children under 5 years of age whose parent or parents both work, are cared for by relatives. Some 24% are cared for in a family day-care home, and 23% in child day-care centers. The remaining 6% of these children are cared for by baby sitters.

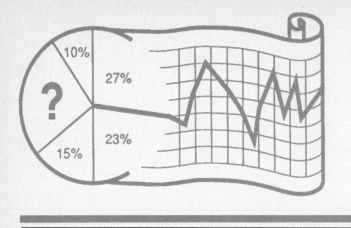

## How Much Do You Already Know?

**Choose the correct completion for each statement. If you are not sure about an answer, do not guess.**

1. Pictographs

   ☐ a. are used to show trends.

   ☐ b. use pictures or symbols to present statistical information.

   ☐ c. are the most accurate of all graphs.

2. Like tables and charts, pictographs arrange information in

   ☐ a. lines.

   ☐ b. bars.

   ☐ c. columns and rows.

3. In a pictograph in which ▮ stands for 10 people, ▎stands for

   ☐ a. 5 people.

   ☐ b. 15 people.

   ☐ c. 20 people.

4. You can expect every pictograph to have a

   ☐ a. footnote to explain the title.

   ☐ b. date telling when the graph was drawn.

   ☐ c. key to explain the symbols used.

**Check your answers on page 149.**

# Pictographs

**7**

**Pictographs** use simple pictures or symbols to present statistical information. Although pictographs are not as precise as tables or graphs, they convey the same information in a way that makes it very easy to understand the data and make immediate comparisons.

## TABLES VS. PICTOGRAPHS

Figure 7.1 is a table showing the number of motor vehicles produced in several countries in 1989.

| Motor Vehicle Production 1989 | |
|---|---|
| Japan | 13,000,000 |
| United States | 11,000,000 |
| Germany | 5,000,000 |
| France | 4,000,000 |
| Italy | 2,000,000 |
| Spain | 2,000,000 |
| South Korea | 1,000,000 |
| Sweden | 500,000 |

**Figure 7.1  Table Showing Motor Vehicle Production**

The table shows that the leading producer of motor vehicles in 1989 was Japan.

How many motor vehicles did that country produce?_____

According to the table, Japan produced 13,000,000 motor vehicles in 1989.

The table lists manufacturers according to the number of cars they produced. The largest car maker is at the top. The information is presented in an orderly fashion and is accurate. But it takes some studying to see the differences in production among the countries.

For instance, how many more motor vehicles than South Korea did Italy produce? _____

According to the table, in 1989 Italy produced 1 million more motor vehicles than South Korea (2,000,000 − 1,000,000 = 1,000,000).

The same information shown in this table could also be presented in a pictograph (see Figure 7.2). Now the difference in production among countries is readily apparent, thanks to the use of symbols instead of figures.

Motor Vehicle Production*
1989

*trucks, buses, and cars
**East and West Germany combined production
(East: 200,000. West: 4.8 million)

KEY: = 1 million motor vehicles

Source: Motor Vehicle Manufacturer's Association.

**Figure 7.2  Pictograph Showing Motor Vehicle Production**

## PARTS OF A PICTOGRAPH

The most obvious part of any pictograph are the pictures, or symbols, themselves. Just as a line identifies a line graph, and a bar a bar graph, the symbols used tell us this is a pictograph.

In Figure 7.2 the symbol  represents automobiles. In another pictograph, the symbol might represent people, or the symbol could mean barrels of oil. Whatever the symbol, it will most often look like the subject the pictograph is presenting information about. If you have any doubts about what that subject is, look at the title of the graph.

What is the title of the pictograph in Figure 7.2? _____

_____

The title of the graph is Motor Vehicle Production 1989.

Notice that the title has an asterisk (*) following it. The asterisk calls attention to a footnote—a note at the bottom of the graph explaining something about the title.

 What information about the title does the footnote provide? _____

_____

The footnote tells us that the title—Motor Vehicle Production 1989—includes cars, trucks, and buses, not just cars.

Besides the title, a pictograph may have a source. The source, usually found at the bottom of the graph, tells where the information in the graph came from.

What is the source of information in the graph in Figure 7.2? _____

_____

The source of information is the Motor Vehicle Manufacturer's Association.

Like other graphs and tables, pictographs present information in an orderly way. Here it is organized into columns and rows. Columns are always read vertically. The column in Figure 7.2 appears at the left of the graph.

What does the column in Figure 7.2 contain? _____

_____

The column lists the names of the countries whose motor vehicle production is measured on the graph.

Rows, which are read from left to right, contain the symbols. Each row in the graph in Figure 7.2 stands for a certain number of motor vehicles.

Finally, every pictograph will have a key, or **legend**. The key will explain the value of each symbol in the graph. The key may appear within the pictograph itself, or it may be at the bottom of the graph.

Find the key for the pictograph in Figure 7.2. How many motor vehicles does

each 🚗 represent? _____

Each 🚗 represents 1 million motor vehicles.

## READING A PICTOGRAPH

In reading any pictograph, remember that you must read down the column and across the rows. Suppose, for example, you want to know how many motor vehicles were produced in France in 1989.

First, look down the column at the left until you find France. Then count the number of symbols in the row following France.

 How many  symbols are in this row?_____

Following France there are four  symbols.

The key tells us that each  represents 1 million motor vehicles. Since the row labeled France includes four of these symbols, you must multiply to find the total number of motor vehicles.

 What is that number?_____

In 1989, France manufactured 4 million motor vehicles.

Now look down the column until you find Germany. Notice that two asterisks (**) follow this word. To find what these asterisks refer to, look at the footnotes at the bottom of the graph.

 What does the footnote following the two asterisks tell us? _____

_____

The footnote tells us the total number of motor vehicles produced in Germany includes those made in East Germany (200,000 vehicles) and those made in West Germany (4,800,000 vehicles).

 Now look for Sweden. This country produced fewer than 1 million motor

vehicles in 1989. How can you tell? _____

_____

Since there is less than one complete  symbol in the row for Sweden, it means that country produced fewer than 1 million cars.

To find how many motor vehicles were produced in Sweden, you will have to estimate. The picture in the row for Sweden shows half of a symbol  . You learned that  represents 1 million motor vehicles.

 How many motor vehicles did Sweden produce in 1989? _____

# **P**ictographs

 **Pictographs** use simple pictures or symbols to present statistical information. Although pictographs are not as precise as tables or graphs, they convey the same information in a way that makes it very easy to understand the data and make immediate comparisons.

## TABLES VS. PICTOGRAPHS

Figure 7.1 is a table showing the number of motor vehicles produced in several countries in 1989.

| Motor Vehicle Production 1989 | |
| --- | --- |
| Japan | 13,000,000 |
| United States | 11,000,000 |
| Germany | 5,000,000 |
| France | 4,000,000 |
| Italy | 2,000,000 |
| Spain | 2,000,000 |
| South Korea | 1,000,000 |
| Sweden | 500,000 |

**Figure 7.1  Table Showing Motor Vehicle Production**

The table shows that the leading producer of motor vehicles in 1989 was Japan.

How many motor vehicles did that country produce?_____

According to the table, Japan produced 13,000,000 motor vehicles in 1989.

The table lists manufacturers according to the number of cars they produced. The largest car maker is at the top. The information is presented in an orderly fashion and is accurate. But it takes some studying to see the differences in production among the countries.

For instance, how many more motor vehicles than South Korea did Italy produce?_____

According to the table, in 1989 Italy produced 1 million more motor vehicles than South Korea (2,000,000 − 1,000,000 = 1,000,000).

The same information shown in this table could also be presented in a pictograph (see Figure 7.2). Now the difference in production among countries is readily apparent, thanks to the use of symbols instead of figures.

Motor Vehicle Production*
1989

| | |
|---|---|
| Japan | 🚗🚗🚗🚗🚗🚗🚗🚗🚗🚗🚗🚗🚗🚗 |
| United States | 🚗🚗🚗🚗🚗🚗🚗🚗🚗🚗🚗 |
| Germany** | 🚗🚗🚗🚗🚗 |
| France | 🚗🚗🚗🚗🚗 |
| Italy | 🚗🚗 |
| Spain | 🚗🚗 |
| South Korea | 🚗 |
| Sweden | 🚗 |

*trucks, buses, and cars
**East and West Germany combined production
  (East: 200,000. West: 4.8 million)

KEY: 🚗 = 1 million motor vehicles

Source: Motor Vehicle Manufacturer's Association.

**Figure 7.2  Pictograph Showing Motor Vehicle Production**

## PARTS OF A PICTOGRAPH

The most obvious part of any pictograph are the pictures, or symbols, themselves. Just as a line identifies a line graph, and a bar a bar graph, the symbols used tell us this is a pictograph.

In Figure 7.2 the symbol 🚗 represents automobiles. In another pictograph, the symbol might represent people, or the symbol 🛢 could mean barrels of oil. Whatever the symbol, it will most often look like the subject the pictograph is presenting information about. If you have any doubts about what that subject is, look at the title of the graph.

What is the title of the pictograph in Figure 7.2? _____

_____

The title of the graph is Motor Vehicle Production 1989.

Sweden produced 500,000 motor vehicles that year. (If represents 1 million motor vehicles, then half a symbol represents half of a million, or 500,000.)

## WARMUP

**Use the graph in Figure 7.2 to answer these questions.**

1. How many motor vehicles were produced in the United States in 1989? _____

2. What two countries produced 2 million motor vehicles each in 1989? _____

   _____

3. How many motor vehicles in all did the United States and Japan produce in 1989?

   _____

4. Other than the United States and Japan, how many motor vehicles did the rest of the

   countries shown in the graph produce? _____

5. How many more motor vehicles did the United States and Japan produce than all the

   other countries put together? _____

**Check your answers on page 149.**

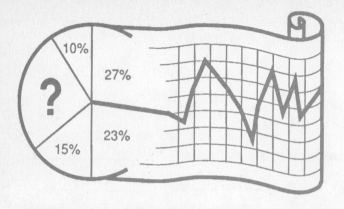

## How Carefully Did You Read?

**A. Choose the correct completion for each statement.**

1. A pictograph is different from other graphs because it has

   ☐ a. title.
   ☐ b. a source.
   ☐ c. pictures or symbols.

2. In a pictograph, the symbol $ might well stand for

   ☐ a. gasoline consumption.
   ☐ b. people employed.
   ☐ c. money.

3. An asterisk (*) on a pictograph usually

   ☐ a. points out any mistakes on the graph.
   ☐ b. calls attention to a footnote.
   ☐ c. identifies the title.

4. In a pictograph, the key

   ☐ a. tells how accurate the graph is.
   ☐ b. explains the pictures or symbols.
   ☐ c. tells where the information on the graph came from.

5. In reading a pictograph, you must read

☐ a. across the columns and down the rows.

☐ b. up the rows and down the columns.

☐ c. down the column and across the rows.

6. In a pictograph in which 🏠 represents 1,000 houses, would represent

☐ a. 500 houses.

☐ b. 1,000 houses.

☐ c. 1,500 houses.

**B. Use the pictograph in Figure 7.3 to complete the sentences.**

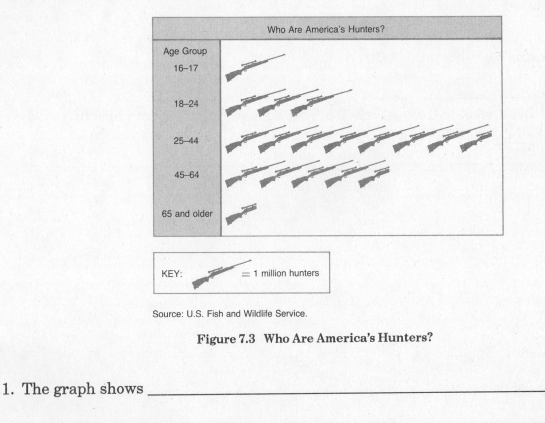

Source: U.S. Fish and Wildlife Service.

**Figure 7.3  Who Are America's Hunters?**

1. The graph shows _____

_____.

2. The source of the information in the graph is _____

_____.

3. According to the graph, the symbol 🔫 represents _____.

4. The age group making up the greatest number of hunters is _____.

5. The age group making up the fewest number of hunters is _____.

6. The difference in numbers between the age group with the most hunters and the age group with the fewest hunters is _____.

7. The age group that contains 3 million hunters is _____.

8. The symbol ⌐ represents _____ hunters.

9. According to the pictograph, the total number of hunters in the United States is

   _____.

10. There are _____ more hunters in the 25 to 44 age group than there are in the 45 to 64 age group.

11. If you owned a sporting goods store that sells hunting gear, you would do best to direct your ads to which age group? _____

**Check your answers on page 149.**

Look back at **How Much Do You Already Know?** on page 70. Did you complete each statement correctly? If not, can you do so now?

Unemployed Workers*
United States

1989

1980

1970

1960

1950

1940

KEY  = 1 million workers

*Workers aged 16 and older

**Figure 7.4  Completing a Pictograph**

ARE YOU READY FOR THE CHALLENGE of making a pictograph from information given you? If you think you are, use the data that follow to complete the graph in Figure 7.4.

In 1940, some 8 million people in the United states were unemployed. Ten years later, in 1950, the number of unemployed workers was 3 million. In both 1960 and 1970 there were 4 million unemployed workers in the country. This figure rose to 7½ million in 1980, and declined to 6½ million in 1989.*

*Source: U.S. Bureau of Labor Statistics.

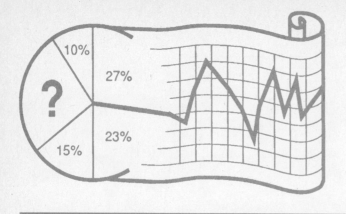

## How Much Do You Already Know?

**Choose the correct completion for each statement. If you are not sure about an answer, do not guess.**

1. A rising line on a graph shows

   ☐ a. an increasing trend.

   ☐ b. a decreasing trend.

   ☐ c. estimated numbers.

2. When reading a graph, to find *why* something happened you must always

   ☐ a. check the source line.

   ☐ b. draw a conclusion.

   ☐ c. find another graph and compare the information.

3. You can accurately draw conclusions

   ☐ a. only from bar graphs.

   ☐ b. only from line graphs.

   ☐ c. from any table or graph.

4. A friend tells you that she will get a new car as soon as her old one breaks down. Two weeks later the old car breaks down. You can conclude

   ☐ a. she will repair the old car before she sells it.

   ☐ b. she will trade in the old car on a new one.

   ☐ c. she will soon be driving a new car.

**Check your answers on page 150.**

# Analyzing Graphs

Graphs present factual information. That is, they tell *what* happened. A line graph, for example, may show the annual temperatures for a certain place. A bar graph may show the growth and decline in the number of soldiers in an army. These are examples of the kind of factual information graphs show. You can find out this kind of information simply by reading it directly from the graph.

In addition, if you do some careful thinking about the information on the graph, or compare the information in one graph with that in another, you may be able to tell *why* something happened. For example, you might be able to make a judgment about why the number of soldiers rose and fell during a certain period of time. Or you might be able to come up with reasons why very few tourists visit a certain city in a certain month.

Finding *why* something happened often means drawing conclusions. Conclusions are judgments based on facts. They are never wild guesses. For example, suppose you read in the newspaper that tickets to a concert you wanted to see were sold out before Wednesday. On Friday, someone you know called to say she had two tickets. Based on the facts, you can conclude she bought the tickets before Wednesday. You could not conclude, however, that she was willing to sell you one of the tickets. There are no facts to support such a conclusion.

Look at the line graph in Figure 8.1. First, let's see *what* the graph shows. Then we'll draw some conclusions based on facts.

## FINDING OUT WHAT HAPPENED

If you aren't sure how to read a line graph, Chapter 4 on page 37 will help you.

 What is the title of the graph? _____

The title is Mean Annual Temperatures, Yakutsk, Russia.

 What is the coldest month in Yakutsk? _____

According to the graph, January is the coldest month. The mean temperature then is −45°F, or forty-five degrees *below* zero!

What is the warmest month in Yakutsk? _____

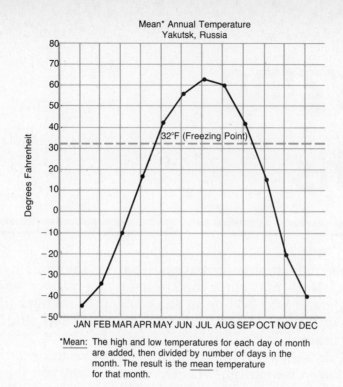

Mean* Annual Temperature
Yakutsk, Russia

32°F (Freezing Point)

Degrees Fahrenheit

JAN FEB MAR APR MAY JUN JUL AUG SEP OCT NOV DEC

*Mean: The high and low temperatures for each day of month
are added, then divided by number of days in the
month. The result is the mean temperature
for that month.

**Figure 8.1  Line Graph Showing Mean Annual Temperature**

Again according to the graph, July is the warmest month. The mean temperature then is 65°F, or 65 degrees *above* zero.

These are just a few of the facts we can learn by reading the graph. Now what are some conclusions we can draw from the graph?

## LEARNING WHY SOMETHING HAPPENS

Yakutsk is located on the Lena River, an important waterway in Russia. During which months would people most likely go fishing and boating on the Lena?

_____

The warmest months—June, July, and August—would probably be the best times to fish or boat on the river.

Ice-skating is one sport the residents of Yakutsk could enjoy during a large part of the year. During which months would they be able to ice-skate?

You can conclude that they could ice-skate from October through April since these are the months during which the mean temperature is below freezing.

## WARMUP A

**Use the graph in Figure 8.1 to answer these questions.**

1. In what months does the temperature in Yakutsk remain above freezing? _____

_____

2. What two months have the same mean temperature? _____

_____

3. How many degrees difference in temperature is there between the coldest month and the warmest month in Yakutsk? _____

4. How much does the mean temperature fall between September and December?

_____

5. Would you say a year-round outdoor job in Yakutsk would be easy or hard to perform?

_____

**Check your answers on page 150.**

## INTERPRETING A BAR GRAPH

Look at the bar graph in Figure 8.2. If you aren't sure how to read a bar graph, Chapter 3 on page 21 will help you.

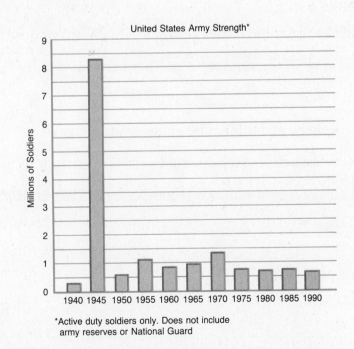

Figure 8.2 **Bar Graph of United States Army Strength**

What does the graph in Figure 8.2 show? _____

_____

The graph shows how many soldiers were on active duty in the United States Army during the years 1940 through 1990.

 What does the scale on the left of the graph measure? _____

_____

This scale measures the number of soldiers on active duty. Each tick mark (–) represents 500,000 soldiers.

 What is the source of information in the graph? _____

_____

The information in the graph came from the Department of the Army.

## WARMUP B

**Use the graph in Figure 8.2 to answer the questions.**

1. In what year did the army have the fewest number of soldiers on active duty?

   _____

2. In what year did the army have the most soldiers on active duty? _____

3. What is the difference in the number of soldiers on active duty in 1940 and the number

   on active duty in 1945? _____

4. In what year were there about 900,000 soldiers on active duty? _____

5. More soldiers serve on active duty during times of war than in peacetime. Knowing

   that, during which two years would you conclude that the United States was at war?

   _____ and _____ .

**Check your answers on page 150.**

## COMPARING TWO GRAPHS

Figure 8.3 includes two bar graphs. The one on the top shows annual per capita income (average income per person regardless of age) in each of four countries.

 What are those countries? _____ _____

_____ _____

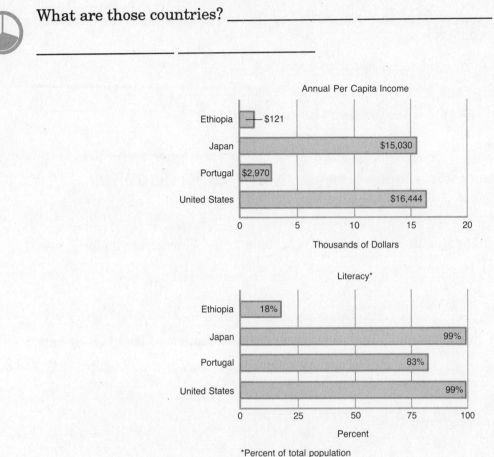

Figure 8.3  Comparing Two Bar Graphs

The four countries represented on the graph are Ethiopia, on the continent of Africa; Japan, a part of Asia; Portugal, a part of Europe; and the United States, a part of North America.

 What does the graph on the bottom show? _____

_____

The graph on the bottom shows the percent of the population in those countries that is literate—that is, the percentage of people who can read and write at a lower elementary school level.

 What is the per capita income of the United States? _____

According to the graph, the per capita income of the United States is $16,444.

 What percent of the United States population is literate? _____

The graph shows that 99% of the people in the United States can read and write at a lower elementary school level.

**WARMUP C**

**Use the graphs in Figure 8.3 to answer the questions.**

1. Which country shows the lowest per capita income? _____

2. What is the difference in per capita income between Portugal and Japan?

   _____

3. What percent of the people in Portugal can read at a lower elementary school level?

   _____

4. Which country has an 18% literacy rate? _____

5. What link can you find between a country's per capita income and its literacy rate?

   _____

**Check your answers on page 150.**

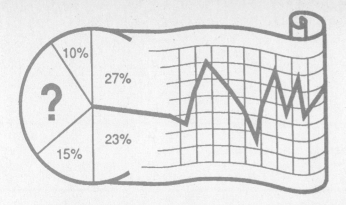

# How Carefully Did You Read?

**A. Choose the correct completion for each statement.**

1. Graphs present

   ☐ a. factual information.

   ☐ b. estimated information.

   ☐ c. conclusions.

2. When reading a graph, to find out *why* something happened you must

   ☐ a. make a guess.

   ☐ b. check the source listed at the bottom.

   ☐ c. draw conclusions.

3. If Sam usually eats out only on his birthday, and he is eating out today, you can conclude that

   ☐ a. his friends like Sam.

   ☐ b. Sam has some extra money.

   ☐ c. today is Sam's birthday.

4. In the term *mean annual temperature*, the word *mean* indicates the

   ☐ a. lowest temperature of the year.

   ☐ b. highest temperature of the year.

   ☐ c. temperature halfway between the low and the high.

5. In a country whose mean annual temperature is 82°F, you would expect to find

☐ a. cold winters.

☐ b. warm winters.

☐ c. lots of rain.

**B. Use the graphs in Figure 8.4 to complete the sentences.**

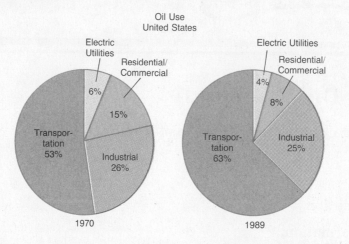

**Figure 8.4  Comparing Two Circle Graphs**

1. The graph on the left shows _____.

2. The graph on the right shows _____.

3. In 1970 the United States used _____ of its oil on transportation.

4. In 1989 the United States used 25% of its oil on _____.

5. In 1970 the United States consumed _____ more oil on

   residential/commercial uses than on electric utilities.

6. Both graphs show the United States uses most of its oil for _____.

7. Both graphs show the United States uses the smallest amount of oil on _____.

8. In 1989 residential/commercial and electric utilities together accounted for

   _____ of the oil consumed in the United States.

9. In what areas did the United States use less oil in 1989 than in 1970? _____

10. The consumption of oil increased by 10% in the area of _____ between

    1970 and 1989.

**Check your answers on page 150.**

Look back at **How Much Do You Already Know?** on page 80. Did you complete each statement correctly? If not, can you do so now?

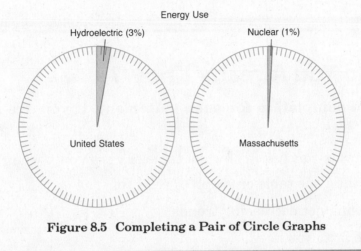

**Figure 8.5 Completing a Pair of Circle Graphs**

**ARE YOU READY FOR THE CHALLENGE** of making a pair of circle graphs? If you think you are, use the data that follow to complete the graphs in Figure 8.5.

If you are not sure how to construct a circle graph, Chapter 6 on page 59 will help you.

In a recent year the United States used the following kinds of energy: oil, 43%; coal, 24%; natural gas, 23%; nuclear power, 7%; hydroelectric power, 3%.

In the same year, Massachusetts consumed these kinds of energy: oil, 69%; natural gas, 18%; coal, 9%; hydroelectric power, 3%; nuclear power, 1%.*

*Source: U.S. Department of Energy.

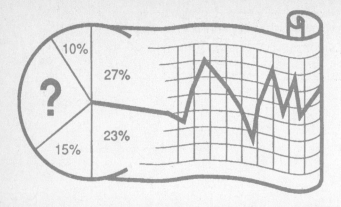

## How Much Do You Already Know?

Choose the correct completion for each statement. If you are not sure about an answer, do not guess.

1. You can draw accurate conclusions from a table or graph only when you
   - ☐ a. have more than one table or graph to refer to.
   - ☐ b. note increasing not decreasing trends.
   - ☐ c. have all the facts you need.

2. Bar graphs can distort, or slant, information by
   - ☐ a. giving false statistics.
   - ☐ b. the way the bars are drawn.
   - ☐ c. making the graphs hard to understand.

3. Line graphs can distort information if
   - ☐ a. different scales are used on similar graphs.
   - ☐ b. rising trends are pictured.
   - ☐ c. broken instead of solid lines are used to show trends.

4. On a pictograph in which the symbol 🏠 stands for 1,000 houses, the most accurate way of showing 2,000 houses is

   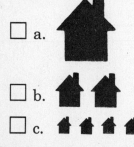

   - ☐ a.
   - ☐ b.
   - ☐ c.

Check your answers on page 150.

# Seeing Is (Not Always) Believing

9

## MISLEADING TABLES

The old saying "seeing is believing" does not always hold true when reading tables and graphs. Sometimes the table may show only part of the information needed to draw a logical conclusion. Look at the table in Figure 9.1.

| Deaths in the United States | |
|---|---|
| Year | Number of deaths |
| 1955 | 1,500,000 |
| 1960 | 1,700,000 |
| 1965 | 1,800,000 |
| 1970 | 1,900,000 |
| 1975 | 1,900,000 |
| 1980 | 2,000,000 |
| 1985 | 2,000,000 |
| 1989 | 2,100,000 |
| Source: National Center for Health Statistics. | |

Figure 9.1  An Example of a Misleading Table

Based on information in the table, what trend do you see in the annual number of deaths in the United States? _____

_____

The table shows the number of deaths has been increasing over the years.

Based on the increasing number of deaths, what conclusion might you draw from the table? _____

_____

One conclusion is that life in the United States seems to be getting more dangerous or unhealthy as the years go by.

But of course this is not true. It only seems true because the table lacks one important piece of information. Can you think what that might be? _____

_____

The table does not include the total population of the United States for each of the years shown on the graph.

In fact, between 1955 and 1989 the population grew by some 37 million people—from 190 million to about 227 million. What accounts for the increase in deaths is the increase in the number of people. (Indeed, if you were to check the death *rate*—the number of deaths per 100,000 people—for the same years, you would see a *decrease*.)

When reading tables, do not jump to conclusions. Make sure you have enough facts at hand to draw logical conclusions.

## DISTORTED BAR GRAPHS

The way a particular bar graph is drawn can distort, or slant, the information it shows. Look at the bar graphs in Figure 9.2.

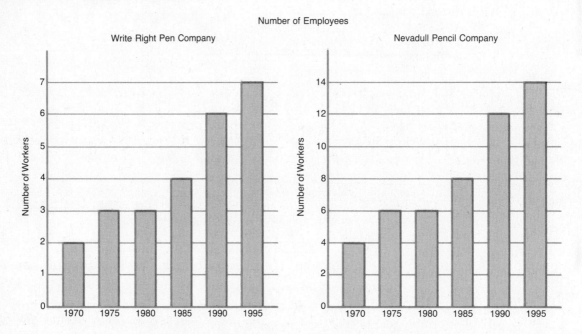

Figure 9.2  An Example of Distorted Bar Graphs

 What does the graph on the left show? _____

_____

The graph on the left shows the number of employees at the Write Right Pen Company for the years 1970 to 1995.

 What does the graph on the right show? _____

_____

The graph on the right shows the number of employees at the Nevadull Pencil Company for the years 1970 to 1995.

At first glance both companies appear to have had the same number of employees during the years shown. After all, the bars in each graph are the same height.

 How many employees did the Write Right Pen Company have in 1985?

_____

According to the graph, the pen company had 4 employees in 1985.

 How many employees did the Nevadull Pencil Company have in 1985?

_____

Again according to the graph, the pencil company had 8 employees in 1985.
How can the numbers differ when the bars are of equal height? The reason is the vertical scales are different on the two graphs.

 What does the vertical scale for the pen company graph show? _____

_____

This vertical scale shows the number of employees, from zero to seven, in increments of one.

 Now look at the vertical scale on the graph showing employees of the pencil

company. What does this scale show? _____

_____

This scale shows the number of workers, from zero to fourteen, in increments of two.
The result is that a bar of any given height on the pencil company graph has twice the value of a bar of the same height on the pen company graph. Had the two graphs used the same vertical scale, they would have looked like the graph in Figure 9.3.
Now the difference in number of employees is quite clear.
When comparing information from two bar graphs, be certain both graphs use the same vertical scales.

## DISTORTED LINE GRAPHS

The way a line graph is drawn can distort our understanding of the information it presents. Look at the two line graphs in Figure 9.4.
Both graphs present the same information—growth in sales for the Aeromodelers Company from 1990 to 1995.

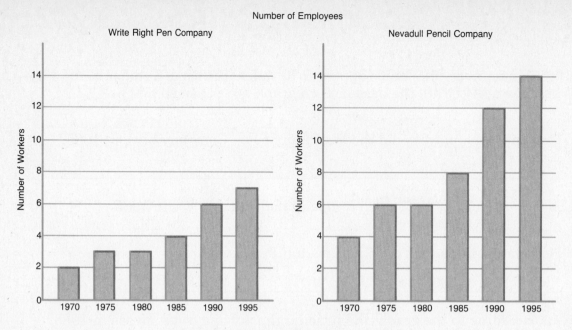

Figure 9.3  Comparing Two Bar Graphs

 What trend in sales does each graph show? _____

_____

Both graphs show a rising trend; that is, increasing sales, for the years shown on the graph.

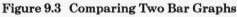 What was the total dollar amount of sales in 1990? _____

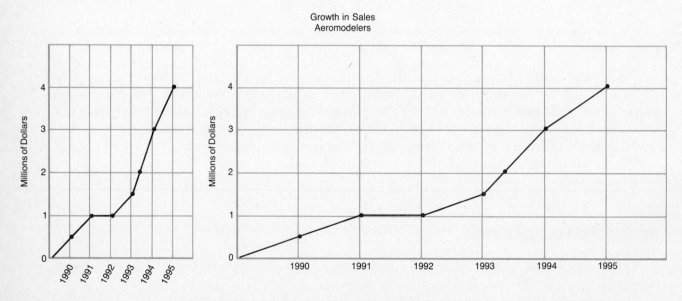

Figure 9.4  An Example of Distorted Line Graphs

Both graphs show sales of $500,000 in 1990.

What was the total dollar amount of sales in 1995? _____

In 1995 Aeromodelers reached sales of $4 million.

Although both graphs contain the same statistical information, one shows a more dramatic increase in sales. Which graph is this? _____

The line graph on the left shows a far sharper rise in sales.

The reason for this can be found by looking at the horizontal scale (the one that runs across the bottom) of each graph. Note the space between the years on each horizontal scale. The scale on the graph at the left has the years very close together. The result is the sharply rising sales line.

On the other hand, the graph on the right has stretched out the space between the years on its horizontal scale. This results in a somewhat flattened line, which suggests the rise in sales was less dramatic.

When reading a line graph, do not be mislead by the relative steepness of the line, whether it is rising or falling. Make your final judgments based on the numerical values the line represents, not on the way the line is drawn.

## DISTORTED CIRCLE GRAPHS

Figure 9.5 includes two circle graphs that present the same information.

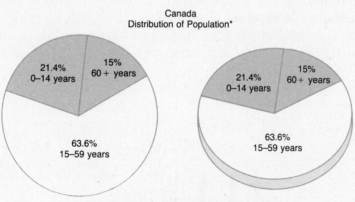

*total population estimated 26,527,000 in 1990

**Figure 9.5  Comparing Two Circle Graphs**

What does each circle graph show? _____

_____

Each graph shows the population distribution of Canada by age group.

⊘ What age group makes up the largest part of the population of Canada?

_____

The graph shows that those between 15 and 59 years of age account for 63.6 percent of Canada's population.

⊘ What part of that country's population is made up of people over 60 years of

age? _____

People over 60 years of age account for 15 percent of Canada's population.

Again, note that both circle graphs present the same information. The graph on the left, however, gives a more accurate picture. In an attempt to make the graph on the right more artistically appealing, the artist tipped it slightly, changing the circular shape to an oval shape. Doing this resulted in making the pie-shaped segment that represents the 15-to-59 year age group (the one that is closest when reading the graph) appear larger than it should be. At the same time, the other two pie-shaped segments (those which are farther away) appear smaller than they should be.

When confronted with circle graphs drawn like the one on the right in Figure 9.5, be alert to distortions of pie-shaped segments caused by the tipping of the graph.

## DISTORTED PICTOGRAPHS

Pictographs use pictures or symbols to present statistical information. Because they use pictures or symbols, pictographs are always less accurate than bar or line graphs. It is especially important in reading any pictograph to draw conclusions based on the statistical information, not the pictures or symbols. Use the pictures only to make quick comparisons.

Look at the pictograph in Figure 9.6. It shows the population of Cypress Gardens, Florida, in 1970 and in 1980, a time of great growth. Figure 9.6 shows one way to present this information.

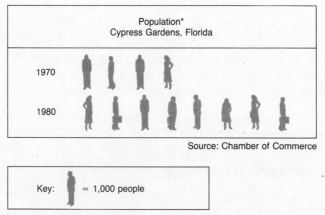

Figure 9.6  An Example of a Pictograph

How many people lived in Cypress Gardens in 1970? _____

The pictograph shows that 4,000 people lived in Cypress Gardens in 1970.

Ten years later how many people were living there? _____

By 1980, the population had increased to 8,000 people.

What was the increase in the number of people between 1970 and 1980?

_____

The population increased by 4,000 people. Put another way, the population doubled in the ten years from 1970 to 1980.

Another pictograph might try to convey the same information as shown in Figure 9.7.

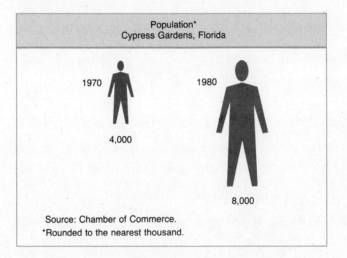

**Figure 9.7  An Example of a Distorted Pictograph**

Drawing a pictograph with symbols like these results in a double distortion. At first glance it looks as if the people themselves grew in size rather than in numbers. Also, the symbol on the right (representing the 1980 population) is twice the size of the 1970 symbol *in all ways*. In other words, it is twice as tall and twice as wide. The result is, that the symbol representing 1980 is *four times* the size of the 1970 one, whereas in reality the population only doubled.

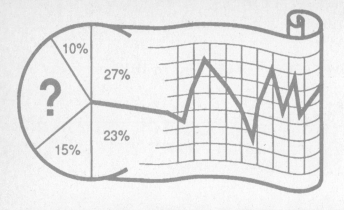

## How Carefully Did You Read?

**A. Choose the correct completion for each statement.**

1. Jumping to conclusions means

   ☐ a. making a judgment based only on bar graphs.

   ☐ b. making a judgment based only on circle graphs.

   ☐ c. making a judgment without having all the facts at hand.

2. When comparing bar graphs be sure

   ☐ a. the bars in each are of the same height or length.

   ☐ b. each graph has the same number of bars.

   ☐ c. the vertical scale on each graph has the same numerical values.

3. A horizontal scale

   ☐ a. runs across the bottom of a graph.

   ☐ b. runs up and down on a graph.

   ☐ c. may be drawn with or without numerical values.

4. A steeply rising or falling line on a line graph is

   ☐ a. less accurate than one that is somewhat flattened.

   ☐ b. more accurate than one that is somewhat flattened.

   ☐ c. neither more nor less accurate than one that is somewhat flattened.

5. A circle graph drawn like this is likely to

☐ a. distort the apparent sizes of the segments.
☐ b. be hard to read.
☐ c. take longer to draw.

6. On a pictograph in which ✈ stands for 100 airplanes, the symbol might seem to stand for

☐ a. 100 airplanes.
☐ b. 1,000 airplanes.
☐ c. 400 airplanes.

**B. Use the graphs in Figure 9.8 to complete the sentences.**

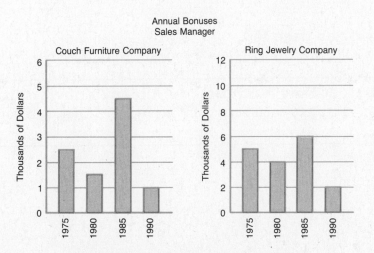

Figure 9.8  Analizing Two Bar Graphs

1. The graph on the left presents statistical information for the _____.

2. The graph on the right presents statistical information for the _____.

3. In 1980 the annual bonus for the sales manager of the Couch Furniture Company amounted to _____.

4. In 1980 the annual bonus for the sales manager of the Ring Jewelry Company amounted to _____.

5. The annual bonus for the furniture company's sales manager reached $4,500 in the year _____.

6. In that same year the annual bonus for the jewelry company's sales manager reached

   _____.

7. The smallest bonus the sales manager of the furniture company earned was

   _____ in the year _____.

8. In the year the furniture company's sales manager earned a bonus of $1,000, the

   jewelry company's sales manager earned a bonus of _____.

9. The largest bonus the sales manager of the jewelry company earned was

   _____ in the year _____.

10. Over the years shown on the graphs, the sales manager of the jewelry company

    earned _____ more than the sales manager of the furniture company

    earned. (You must add, then subtract to find the answer.)

**Check your answers on page 150.**

> Look back at **How Much Do You Already Know?** on page 90.
> Did you complete each statement correctly? If not, can you do so
> now?

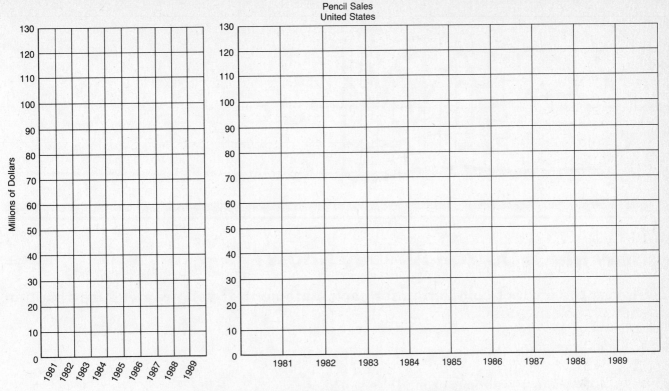

Pencil Sales
United States

**Figure 9.9  Completing a Pair of Line Graphs**

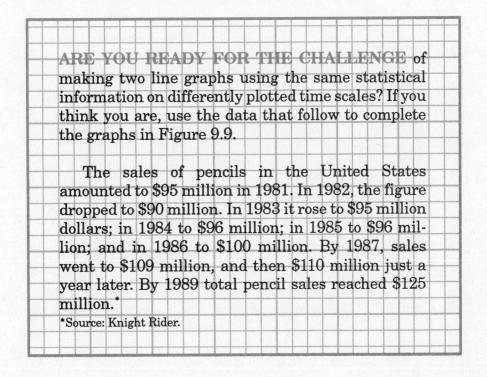

ARE YOU READY FOR THE CHALLENGE of making two line graphs using the same statistical information on differently plotted time scales? If you think you are, use the data that follow to complete the graphs in Figure 9.9.

The sales of pencils in the United States amounted to $95 million in 1981. In 1982, the figure dropped to $90 million. In 1983 it rose to $95 million dollars; in 1984 to $96 million; in 1985 to $96 million; and in 1986 to $100 million. By 1987, sales went to $109 million, and then $110 million just a year later. By 1989 total pencil sales reached $125 million.*

*Source: Knight Rider.

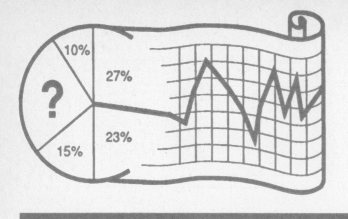

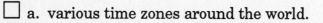

## How Much Do You Already Know?

**Choose the correct completion for each statement. If you are not sure about an answer, do not guess.**

1. Time lines show

   ☐ a. various time zones around the world.

   ☐ b. when certain events took place.

   ☐ c. rising or falling trends.

2. The terms B.C. and A.D. refer to time

   ☐ a. before and after noon.

   ☐ b. in various parts of the world.

   ☐ c. before and after the birth of Christ.

3. An event that happened in the year 100 B.C. occurred

   ☐ a. before the year 200 B.C.

   ☐ b. after the year 99 B.C.

   ☐ c. before the year 100 A.D.

4. An event that occurred in the seventeenth century happened between the years

   ☐ a. 1601 and 1700.

   ☐ b. 1701 and 1800.

   ☐ c. 1801 and 1900.

**Check your answers on page 151.**

# Time Lines

# 10

## PURPOSE OF A TIME LINE

**Time lines** provide an easy and quick way to illustrate information about when events happened.

Look at this passage about the history of Canada.

Most historians recognize Jacques Cartier as the founder of what is now Canada. This claim is based on his discovery of the Gulf of St. Lawrence in 1534. Yet it is known that John Cabot, an Italian explorer working for the English, sailed by Newfoundland 37 years before Cartier, in 1497. Still, it was the French who settled the region. They founded Quebec City in 1608 and Montreal in 1642. Twenty-one years later, in 1663, the French declared this vast area of the New World a colony, and named it New France.

The passage is filled with information about certain events in the history of Canada. In order to sort out the events, however, you have to read through the entire passage, putting them in the correct sequence, or order. Did Cabot visit Canada before or after Cartier? Was Montreal settled before or after the colony was named New France?

The same information presented in the form of a time line might look like Figure 10.1.

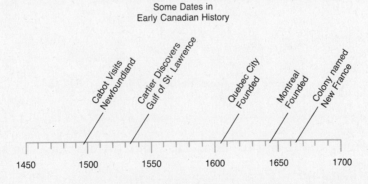

Source: World Almanac and Book of Facts, 1991.

**Figure 10.1  An Example of a Time Line**

Now the sequence of events in the passage is clear and easier to read. The time line shows which event happened first, which happened next, and so on. The line also shows how much time passed between one event and another.

## PARTS OF A TIME LINE

A time line is simply a line, drawn horizontally or vertically, with a series of measures of time on it. (See Figure 10.2.)

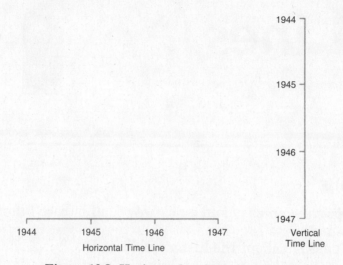

Figure 10.2  Horizontal and Vertical Time Lines

The measures of time are often years, but they may be months, weeks, days, hours, minutes—or even seconds. The measure of time will appear either below the line (on a horizontally drawn time line), or to the left of the line (on a vertically drawn time line). Time periods will be set off by tick marks (–).

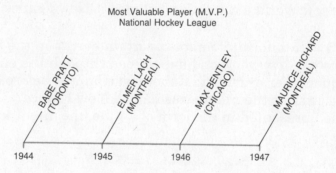

Source: World Almanac and Book of Facts, 1991.

Figure 10.3  Most Valuable Players (NHL)

 Look at the time line in Figure 10.3. How much time does the space between tick marks represent? _____

The space between tick marks represents one year.

Like most other visual aids (graphs, for example), time lines have a time which tells what kind of information is shown.

**104    Chapter 10**

What does the time line in Figure 10.3 show? _____

_____

This time line shows the most valuable players in the National Hockey League for the years 1944 through 1947.

 Finally, a time line may have a source for the information on it. What is the

source of the information on the time line in Figure 10.3? _____

The *World Almanac and Book of Facts* is the source of information on the time line.

## READING A TIME LINE

In reading any time line, first look for the title to learn what the time line shows. Then look at the measures of time to see how much time the space between tick marks represents. If the time line is drawn horizontally, read from left to right. If the time line is drawn vertically, read from top to bottom.

## WARMUP A

**Use the time line in Figure 10.3 to answer the questions.**

1. How many years does the time line span? _____

2. Who won the M.V.P. award in 1944? _____

3. Who won the M.V.P. award in 1947? _____

4. How many different teams are represented by M.V.P.s? _____

5. In what year was Max Bentley named M.V.P.? _____

**Check your answers on page 151.**

## ESTIMATING DATES ON A TIME LINE

In reading a time line, a given date may fall between two measures of time. For example, look back at the time line in Figure 10.1 on page 103. There you will see that each of the events on the line falls between tick marks.

Suppose you wanted to know when Quebec City was founded. You can see the event falls between two measures of time.

 What are those two measures of time?_____

The line shows Quebec City was founded sometime between 1600 and 1650.

Looking more closely at the line, you will see that the event occurred much nearer to 1600 than to 1650. Note that each measure of time is further divided in ten-year increments. The shorter tick marks represent 1610, 1620, 1630, and 1640. A good estimate for the year Quebec City was founded would be 1608.

## WARMUP B

**Use the time line in Figure 10.1 to answer the questions.**

1. How many events does the time line show? _____

2. What is the earliest event shown? _____

3. What is the latest event shown? _____

4. About when was Montreal founded? _____

5. Which event took place in 1534? _____

**Check your answers on page 151.**

### B.C. AND A.D.

Have you ever read any of Aesop's Fables such as "The Fox and the Grapes" or "A Wolf in Sheep's Clothing" ? What do you know about a woman named Cleopatra, who lived in ancient Egypt?

Although Aesop was thought to be a Greek peasant, while Cleopatra was once Queen of Egypt, both have at least one thing in common. Both lived many years before the birth of Christ.

When describing events that took place before the birth of Christ we use the abbreviation B.C. which stands for *before Christ*.

We believe that Aesop lived from about 620 B.C to 564 B.C. Cleopatra was born in 69 B.C. and died in 30 B.C. Every event that occurred before the birth of Christ will carry the abbreviation B.C. after the date.

By way of contrast, the date of every event that occurred after the birth of Christ is considered to be A.D., an abbreviation for the Latin term *Anno Domini*, which means "in the year of the Lord."

Events that occurred after the birth of Christ are most often written without the abbreviation A.D. For example, if we wrote that "Marco Polo left his native Venice in 1271 to start on a perilous journey to China," we know from our study of history that 1271 means 1271 A.D. and not 1271 B.C.

On a time line, B.C. and A.D. dates would look as they do in Figure 10.4.

| 500 B.C. | 400 B.C. | 300 B.C. | 200 B.C. | 100 B.C. | 1 A.D. | 100 A.D. | 200 A.D. | 300 A.D. | 400 A.D. | 500 A.D. |
|---|---|---|---|---|---|---|---|---|---|---|
| | | | | | (Birth of Christ) | | | | | |

**Figure 10.4  Showing B.C. and A.D. on a Time Line**

All dates start from 1 A.D., the year established as that of the birth of Christ.

Starting with 1 A.D. and reading left, do the dates increase or decrease?

_____

B.C. dates increase as they go back in history (toward the left on the time line). A.D. dates, on the other hand, increase as they go forward in history (toward the right on the time line).

Put another way, an event that occurred in 400 B.C. took place *before* an event that happened in 200 B.C. An event that took place in 400 A.D. happened *after* an event that happened in 200 A.D.

Any event that took place in a year before the birth of Christ will always carry the abbreviation B.C. after it. The date of Cleopatra's death is 30 B.C.

## FIGURING ELAPSED TIME

To figure out how many years have passed between two dates, both of which are either B.C. or A.D., you must subtract. Suppose, for example, you wanted to know how many years elapsed between the time Aesop was born and Cleopatra was born. To find out, subtract 69 B.C. from 620 B.C.

$$
\begin{array}{r}
620 \text{ B.C. (year Aesop was born)} \\
- \phantom{0}69 \text{ B.C. (year Cleopatra was born)} \\
\hline
551 \text{ years}
\end{array}
$$

Marco Polo was born in 1254 A.D. while the American poet Phyllis Wheatley was born about 1750 A.D.

How many years elapsed between the births of Marco Polo and Phyllis Wheatley? _____

Phyllis Wheatley was born 496 years after Marco Polo. (Did you remember to subtract?)

$$
\begin{array}{r}
1750 \text{ A.D. (year Wheatley was born)} \\
-1254 \text{ A.D. (year Marco Polo was born)} \\
\hline
496 \text{ years}
\end{array}
$$

Now suppose you want to know how many years elapsed between the births of Cleopatra and Phyllis Wheatley. Remember that Cleopatra was born in 69 B.C. while Phyllis Wheatley was born in 1750 A.D. Because one date is B.C. and the other is A.D., you will have to *add* the BC. date to the A.D. date to find the answer.

$$
\begin{array}{r}
1750 \text{ A.D. (Phyllis Wheatley's birth)} \\
+ \phantom{0}69 \text{ B.C. (Cleopatra's birth)} \\
\hline
1819 \text{ years}
\end{array}
$$

A time line might illustrate the situation in the manner shown in Figure 10.5.

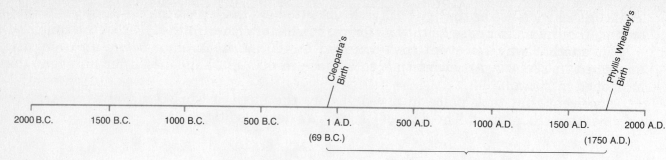

Figure 10.5   Using a Time Line

## WARMUP C

Use the time line in Figure 10.6 to answer the questions.

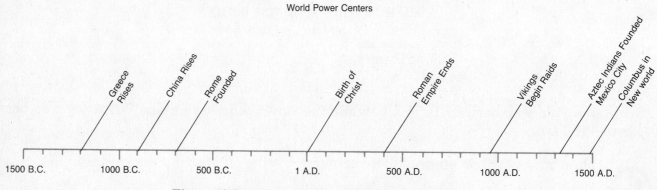

Figure 10.6   A Time LIne of World Power Centers

1. How many years in all does the time line cover? (You must add to find the answer.)

   _____

2. When did the Roman Empire end? _____

3. What important event occurred in 700 B.C.? _____

4. How many years elapsed between the rise of Greece and the founding of Rome?

   _____

5. What important event occurred about 950 A.D.? _____

Check your answers on page 151.

## TALKING ABOUT CENTURIES

In your reading you may have come across statements like "The American Revolution took place in the eighteenth century," or "The Aztec Indians of central Mexico founded what is now Mexico City in the fourteenth century." We also know that we live in the twentieth century, and that soon we will be entering the twenty-first century.

A century is a measure of time equal to 100 years. Rather than giving a precise year, events may be said to have taken place in a certain century, as the previous examples show.

But what years does the eighteenth century or the fourteenth century include? To find out, look at the chart in Figure 10.7.

| Century | Years |
|---------|-------|
| First | 1 A.D. to 100 A.D. |
| Second | 101 to 200 |
| Third | 201 to 300 |
| Fourth | 301 to 400 |
| Fifth | 401 to 500 |
| Sixth | 501 to 600 |
| Fourteenth | 1301 to 1400 |
| Fifteenth | 1401 to 1500 |
| Sixteenth | 1501 to 1600 |
| Seventeenth | 1601 to 1700 |
| Eighteenth | 1701 to 1800 |
| Nineteenth | 1801 to 1900 |
| Twentieth | 1901 to 2000 |
| Twenty-first | 2001 to 2100 |

Figure 10.7  A Chart Showing Centuries of Time

 What years does the fourth century include? _____

The chart in Figure 10.7 shows that the fourth century includes the years 301 to 400. (Again, when dealing with centuries after the birth of Christ, there is no need to use the abbreviation A.D.)

Although the chart breaks off at the sixth century and picks up again at the fourteenth century, perhaps you can figure out what years the missing centuries contain. (Remember: A century is 100 years.)

 What years does the seventh century contain? _____

If you said the seventh century includes the years 601 to 700, you are right.

 What years does the tenth century include? _____

The tenth century includes the years 901 to 1000.

 You may remember that Cleopatra was born in the year 69 B.C. What century was that year a part of?_____

(Remember, when dealing with events before the birth of Christ, whether years or centuries, always include the abbreviation B.C in the dates.)

Cleopatra was born in the first century B. C. (The year 69 B.C. falls between 100 B.C. and 1 B.C.)

## WARMUP D

**Use the chart in Figure 10.7 to answer the questions.**

1. What years does the sixteenth century include? _____

2. What years does the eleventh century include? _____

3. An event that took place in 1775 would have occurred in what century? _____

4. Which of these events took place in the nineteenth century? Texas broke away from

   Mexico in 1835. George Washington was chosen president of the United States in 1789.

   Hawaii became the fiftieth state of the Union in 1959. _____

5. The Greek storyteller Aesop died about 564 B.C. What century was that? _____

**Check your answers on page 151.**

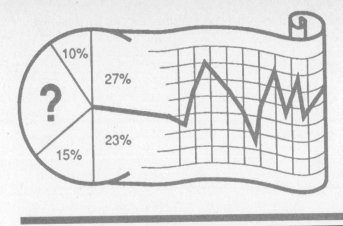

# How Carefully Did You Read?

**A. Choose the correct completion for each statement.**

1. A time line is useful in showing

    ☐ a. only events that occurred in the A.D. period.

    ☐ b. a sequence of events over a period of time.

    ☐ c. trends that occurred over a period of time.

2. Vertical time lines are read from

    ☐ a. left to right.

    ☐ b. top to bottom.

    ☐ c. bottom to top.

3. You must estimate dates on a time line when the events

    ☐ a. are not in any sequence.

    ☐ b. fall between two measures of time on the line.

    ☐ c. occurred in the B.C. period.

4. According to the time line,

    Italy Joins United Nations
    
    1950     1960

    Italy became a member of the United Nations in the year

    ☐ a. 1950.

    ☐ b. 1955.

    ☐ c. 1960.

5. If you read that an event occurred in the year 1929, you can assume that it happened

☐ a. before the birth of Christ.
☐ b. after the birth of Christ.
☐ c. either before or after the birth of Christ.

6. A century contains

☐ a. 1,000 years.
☐ b. 100 years.
☐ c. 10 years.

7. The number of years that elapsed between 600 B.C. and 250 B.C. is

☐ a. 850 years.
☐ b. 450 years.
☐ c. 350 years.

8. The number of years that elapsed between 55 B.C. and 100 A.D. is

☐ a. 45 years.
☐ b. 55 years.
☐ c. 155 years.

**B. Use the time line in Figure 10.8 to complete the statements.**

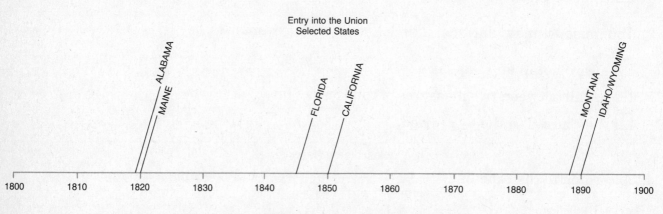

Figure 10.8  Completing a Time Line

1. The time line shows _____.

_____

2. The source of the information on the time line is _____

_____.

3. The time line covers the years from to _____.

4. The number of states admitted to the Union shown on the time line is _____.

5. The state admitted earliest was _____.

6. California was admitted to the Union in _____.

7. The state of _____ was admitted to the Union in 1845.

8. The states of _____ and _____ were both admitted to the Union in the same year.

9. The state of _____ was admitted to the Union in 1819.

10. The state of _____ was admitted to the Union in the year 1889.

11. The time between the admissions of Alabama and Montana was _____ years.

12. Florida was admitted to the Union _____ years after the state of Maine.

**Check your answers on page 151.**

Look back at **How Much Do You Already Know?** on page 102. Did you complete each statement correctly? If not, can you do so now?

50–Homer Club
National League

| | | | | | |
|---|---|---|---|---|---|
| 1930 | 1940 | 1950 | 1960 | 1970 | 1980 |

**Figure 10.9**

**ARE YOU READY FOR THE CHALLENGE** of making a time line from information given you? If you think you are, use the data that follow to fill in the time line in Figure 10.9.

Hitting 50 home runs in a single season is a remarkable feat in baseball. Over the years only six players in the National League have done this. The first was Hack Wilson, who hit 50 in 1930. In 1947 Ralph Kiner and Johnny Mize each hit 51. Three years later Ralph Kiner hit another 54. In 1955 Willie Mays hit 51, and ten years later he hit 52. The latest player to hit more than 50 home runs was George Foster. He hit 52 in 1977.*

*Source: *World Almanac and Book of Facts,* 1991.

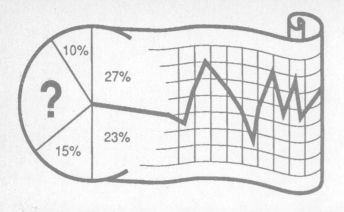

# How Much Do You Already Know?

Choose the correct completion for each statement. If you are not sure about an answer, do not guess.

1. Flowcharts

    ☐ a. illustrate trends.

    ☐ b. show the steps in performing a task or solving a problem.

    ☐ c. arrange information in columns and rows.

2. In a flowchart, oval-shaped figures tell

    ☐ a. where the flowchart begins and ends.

    ☐ b. what decisions have to be made.

    ☐ c. what the next step in a process is.

3. Decision blocks on a flowchart are shown by

    ☐ a. oval-shaped figures.

    ☐ b. diamond-shaped figures.

    ☐ c. rectangles.

4. In reading any flowchart, the first step is to

    ☐ a. see how many steps are involved in the process.

    ☐ b. see which way the arrows run.

    ☐ c. find the start block.

**Check your answers on page 151.**

# Flowcharts

**11**

 In one way flowcharts are like time lines. Both are visual aids that arrange information in a time sequence. They show what happens first, what happens next, and so on.

But flowcharts differ from time lines in one important way. Flowcharts show the steps to follow in performing a task or solving a problem.

For example, take a task such as baking a cake. What steps are involved? They include gathering the ingredients, heating the oven, mixing the ingredients, pouring the batter into a cake pan, and so on.

Figure 11.1 shows how the steps in the task of baking a cake might look on a flowchart.

## PARTS OF A FLOWCHART

You can see that a flowchart consists of a number of differently shaped figures, called blocks, that are connected by arrows. In Figure 11.1 the blocks are ovals ⬭ and rectangles. ▭

The blocks that tell where a flowchart starts and ends are always oval.

 How many rectangles does the flowchart in Figure 11.1 have? _____

The flowchart has eight rectangles.

Rectangles in flowcharts are referred to as operation blocks or process blocks. The information in rectangles tells you exactly what to do.

 Find the first rectangle, or process block, in the flowchart. What does it tell you

to do? _____

The first process block tells you to gather the ingredients.

Note that an arrow leads out of this rectangle to the next, and so on, right down to the

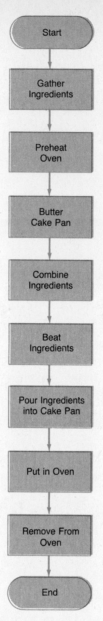

**Figure 11.1 A Flowchart of the Steps in Baking a Cake**

end block, which is oval. Although they are not numbered, the process blocks are arranged in order from the first step to the last step in the task.

 What is the last step? _____

_____

The last step shown on the flowchart is to remove the cake from the oven.

## ADDING DECISION BLOCKS

 What is the seventh process block—the next-to-the-last step in the task of

baking a cake? _____

The seventh step is to bake the cake in the oven.

 Can you think of another step which might come between putting the cake in the oven and removing the cake from the oven? _____

_____

You might consider adding another step, such as testing the cake to see that it is done before taking it from the oven.

Checking to see if the cake is done involves making a decision. If the cake is done, you will want to remove it from the oven. But if it is not done, you will want it to bake some more.

Steps in a task that require a choice are shown on a flowchart by decision blocks. Decision blocks are diamond-shaped figures. They always contain yes or no questions.

Figure 11.2 shows how the flowchart will look with a decision block added.

Notice that the decision block has two arrows leading from it. The arrow labeled yes means the cake is done. It points to the next step in the flowchart, the process block that says "remove from oven."

 Which arrow tells us the cake is not done? _____

The arrow labeled "no" tells us the cake is not yet done.

 Where does the "no" arrow lead? _____

The "no" arrow leads to a process block that tells you to continue baking the cake.

Notice that an arrow leads from the "continue baking" process block back to the "put in oven" block, which completes the loop.

## READING A FLOWCHART

In reading any flowchart, look first for the start block, which will be an oval-shaped figure. On many flowcharts the oval will say "start." On others the oval may include a title, such as "Baking a Cake."

 Next, follow the arrow from the start block to the next block. What kind of figure does the arrow connect the start block to? _____

_____

The arrow connects the start block to a rectangle.

Since the rectangle is a process block, you can expect to be told what to do next.

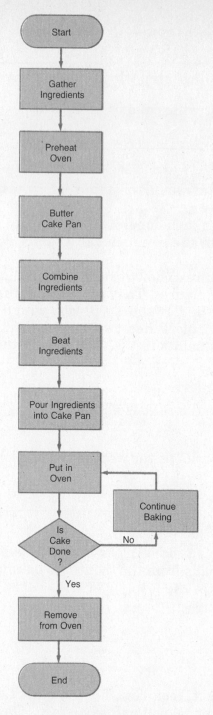

**Figure 11.2  A Flowchart with a Decision Block Added**

What does the first process block tell you to do? _____

_____

The first process block tells you to gather the ingredients.

From this point on, reading a flowchart involves following the arrows from one process block to the next until you reach the end block, which will be another oval-shaped figure. Of course, should the flowchart have decision blocks (identified by their diamond shape), you will need to make a choice based on whether your answer to the question in the decision block is "yes" or "no," before you find your way to the end of the flowchart.

## WARMUP

**Use the flowchart in Figure 11.2 to answer the questions.**

1.  How many steps in the task of baking a cake does the flowchart contain? (Do not include the start and end blocks in your count.) _____

2.  What step follows "combine ingredients?" _____

3.  What does the last process block tell you to do? _____

4.  How many steps are there between "preheat oven" and "pour ingredients into cake pan?" _____

5.  What does the decision block ask you to do? _____

**Check your answers on page 151.**

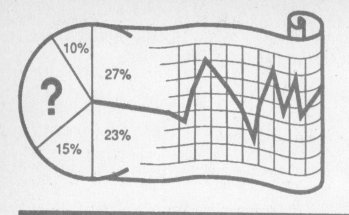

---

## How Carefully Did You Read?

**A. Choose the correct completion for each statement.**

1. Flowcharts are like time lines in that both

   ☐ a. contain instructions.
   ☐ b. use vertical scales.
   ☐ c. arrange information in a time sequence.

2. The flowchart figure that identifies a decision block is the

   ☐ a.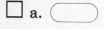

   ☐ b. ▭

   ☐ c. ◇

3. The information in a rectangle tells you

   ☐ a. you must make a decision.
   ☐ b. exactly what to do.
   ☐ c. that the flowchart has ended.

4. An arrow on a flowchart

   ☐ a. points to the next step in the process.
   ☐ b. signals that a decision has to be made.
   ☐ c. answers any questions you may have.

5. An oval-shaped block always signals

☐ a. where the flowchart begins or ends.

☐ b. a "yes" or "no" question.

☐ c. what the next step in the process is.

6. In reading a flowchart, first

☐ a. look for the source line.

☐ b. see how many steps are involved.

☐ c. find the start block.

**B. Use the flowchart in Figure 11.3 to complete the statements. (The flowchart shows the steps involved in leaving the house in the morning.)**

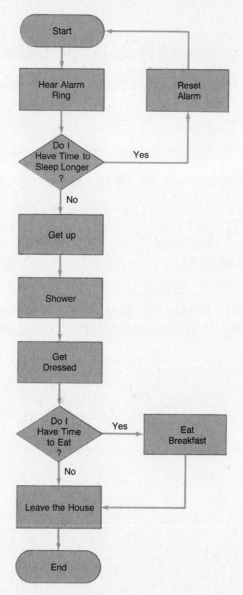

Figure 11.3  Using a Flowchart

1. The first step in the task of getting out of the house is _____

   _____ .

2. The question in the first decision block is _____

   _____ .

3. Assuming that you do have time to sleep longer, the flowchart tells you to _____ .

4. After you get up, the next two process blocks tell you to _____

   and _____ .

5. The second decision block asks the question _____ .

6. Assuming you do not have time to eat breakfast, the flowchart tells you to

   _____ .

7. If you answer "no" to the questions in both decision blocks, the flowchart shows a total

   of _____ steps in the task of leaving the house. (Do not include the start

   and end blocks in your count, but do include decision blocks.)

8. If you answer "yes" to the questions in both decision blocks, the flowchart shows

   _____ steps in the task of leaving the house. (Include all repreated steps

   in your count.)

**Check your answers on page 151.**

Look back at How Much Do You Already Know? on page 116.
Did you complete each statement correctly? If not, can you do so
now?

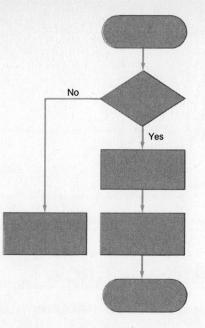

**Figure 11.4   Completing a Flowchart**

ARE YOU READY FOR THE CHALLENGE of completing a flowchart from information given you? If you think you are, arrange the steps in the task of going shopping into the proper sequence. Then enter them in the correct blocks on the flowchart in Figure 11.4.

Stay home
End
Prepare list of items needed
Start
Do I have time?
Go shopping

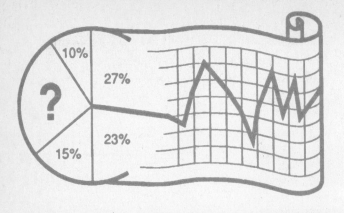

## How Much Do You Already Know?

**Choose the correct completion for each statement. If you are not sure about an answer, do not guess.**

1. A diagram is

   ☐ a. another kind of flowchart.
   ☐ b. found only in a science textbook.
   ☐ c. a kind of picture outline.

2. When a diagram uses letters and numbers to identify the parts shown, it will have a

   ☐ a. key.
   ☐ b. time line.
   ☐ c. source.

3. Which of the following would you *not* expect to be shown as a diagram?

   ☐ a. the parts of a bicycle.
   ☐ b. the seating plan of a theater.
   ☐ c. a photograph of the moon.

4. All diagrams

   ☐ a. tell how to build something.
   ☐ b. tell how something works.
   ☐ c. name important parts.

**Check your ansers on page 151.**

# Diagrams

**Diagrams** are one of the most commonly found, most useful, of all visual aids. By presenting a kind of picture outline, diagrams help make complicated ideas clear and therefore easier to understand. Compare this passage describing a light bulb with the diagram in Figure 12.1.

Incandescent Lamp

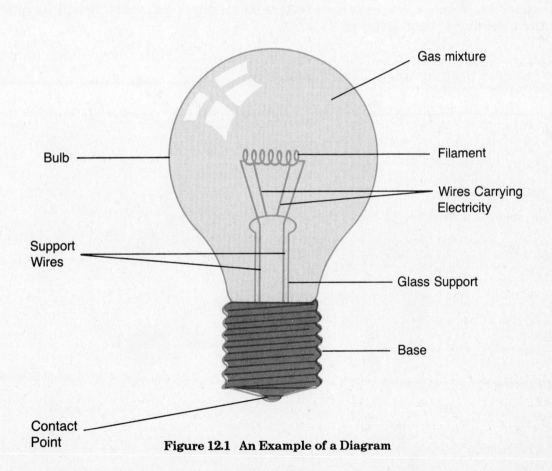

**Figure 12.1 An Example of a Diagram**

The most common source of electric light is the incandescent lamp. Incandescent lamps (the word *incandescent* means "glowing with heat") consist of a filament, a bulb, and a base. The bulb and the base help the filament do its job, which is to produce light.

The filament is a coil of very thin wire. When the lamp switch is turned on, a current of electricity flows through the filament, heating it until it glows and gives off light.

The filament is encased in a bulb, sealing it from air, which would cause the filament to burn up. In place of air, the bulb usually contains a mixture of gases. These help increase the life of the filament.

Both filament and bulb are attached to a brass base. The base seals the entire lamp and provides a way of connecting it to a source of electricity through a contact point at the bottom.

Now look at the diagram in Figure 12.1. It presents much of the same information in a visual way.

The diagram pictures the shape of an incandescent lamp, and the labels point out its major parts. The result is a clear, easy way to understand the visual description of an incandescent lamp.

## DIAGRAMS WITH KEYS

Some diagrams label the various parts with letters or numbers instead of naming the parts themselves. When a diagram uses letters or numbers as labels, it will have a key to explain the meaning of each letter or number.

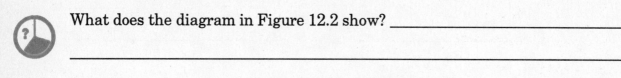 What does the diagram in Figure 12.2 show? _____

_____

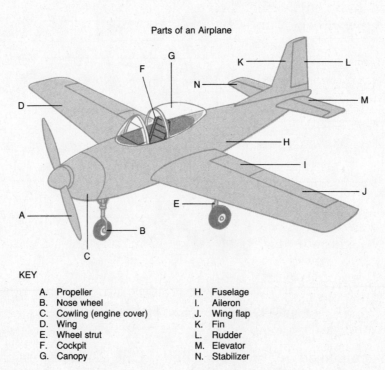

Parts of an Airplane

KEY

| | | | |
|---|---|---|---|
| A. | Propeller | H. | Fuselage |
| B. | Nose wheel | I. | Aileron |
| C. | Cowling (engine cover) | J. | Wing flap |
| D. | Wing | K. | Fin |
| E. | Wheel strut | L. | Rudder |
| F. | Cockpit | M. | Elevator |
| G. | Canopy | N. | Stabilizer |

**Figure 12.2  A Diagram with a Key**

The diagram shows some important parts of an airplane.

Rather than naming the parts, this diagram labels each with a letter. The key with the diagram tells what part of an airplane each letter stands for.

 What part of an airplane does the letter **D** represent? _____

As the key shows, the letter **D** is the label for the airplane's wing.

 What letter is used for the airplane's rudder? _____

The rudder is labeled **L** on the diagram.

Notice that this diagram simply names the parts of an airplane. It does not explain the purpose of any of the parts. It does not show, for example, what function an elevator or a wing flap has in flying the airplane.

This is true of many diagrams. In fact, most diagrams do two things. They (1) show and name all the important parts of something; and (2) show how the parts fit together, or relate.

The diagram in Figure 12.3 shows the floor plan of a library. Instead of words, letters are used to label each part of the library.

 Where can you learn what each letter in the diagram stands for? _____

_____

To find the name of the part each letter stands for, look in the key.

The key includes seventeen items, from the main entrance (**A**) to the restrooms (**Q**).

## READING A DIAGRAM

The first step in reading any diagram is to learn what the diagram shows. Look for a title, which will tell you what the diagram shows.

 What is the title of the diagram in Figure 12.3?_____

_____

The title of this diagram is Library Floor Plan.

A floor plan shows where things in a room or a building are located. Diagrams that are floor plans are drawn so they give what is called a bird's eye view. The view is from above, so you are looking straight down on the interior of the library.

Once you know what the diagram shows, study it carefully, to see what the various parts are. Keep studying the diagram until you are sure you understand it. To see that you do, ask yourself questions from time to time.

 For example, suppose you want to take out a movie video. What letter identifies

the audio-visual room, where movie videos would be kept? _____

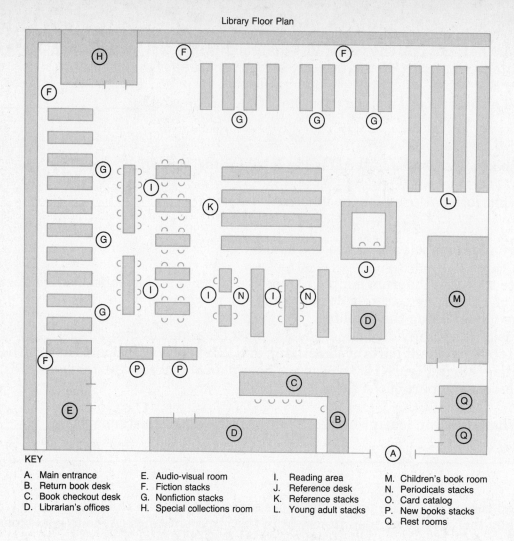

Figure 12.3  A Floor Plan Diagram with Key

The diagram shows the letter **E** stands for the audio-visual room.

To find that room, you would go through the main entrance (**A**), past the book return desk (**B**), and turn left at the book checkout desk (**C**). Then you would go straight across that room to the audio-visual room (**E**).

Now suppose you want to find out where a certain book is located in the library. You already know the title of the book and the name of the author. You also know the book is nonfiction; that is, a book that deals with real people and real events.

What letter identifies the location of the nonfiction stacks, the shelves where

this kind of book can be found? _____

The key tells us the letter **G** indicates areas in the library where nonfiction books are shelved.

Notice, however, that the diagram shows nineteen stacks (or rows of shelves) for nonfiction books. These may hold thousands of books, while you want to find just one.

 What area of the library would you go to learn exactly where the book you want

is shelved? _____

You should go to the card catalog, which lists the location of every book in the library.

 What letter on the diagram identifies the card catalog? _____

The letter **O** stands for the card catalog.

## UNDERSTANDING HOW THE PARTS RELATE

All diagrams name important parts. Most also show how the parts relate to each other. The way the parts relate may not always be spelled out for you, however. You may have to think a bit, to draw conclusions as to how the various parts fit together.

 What do the letters **A**, **B**, and **C** in Figure 12.3 identify? _____
_____

The letter **A** identifies the main entrance to the library. The letter **B** identifies the book return desk, and **C** the book checkout desk.

 Why do you suppose the return and checkout desks are located close to the

main entrance? _____
_____

Having the return and checkout desks close to the main entrance is a convenience for both the librarians and the people using the library.

 What does the letter **N** stand for on the diagram? _____
_____

The letter **N** identifies the periodicals stacks, the area where magazines and newspapers are shelved. The letter **K** points out the reference books.

 What is located between the periodicals stacks and the reference book area?
_____

A reading area is located between the periodicals stacks and the reference book area.

Why is it logical to put a reading area in this location? _____

_____

In most libraries the most recent issues of magazines and newspapers cannot be taken out of the library. Reference books must also be used in the library. The reading area in that location is a convenience for readers of periodicals and reference books.

## WARMUP A

**Use the diagram in Figure 12.3 to answer the questions.**

1. What does the letter **F** identify? _____

2. What letter identifies the stacks that hold books for young adults? _____

3. What does the letter **K** identify? _____

4. What letter identifies the children's book room? _____

5. Fiction books, those that deal with imaginary people and events, are usually arranged

   alphabetically by the author's last name. Facing the main entrance, would you expect

   a book written by Isaac Asimov to be found on the fiction shelves to the left, or on the

   shelves set against the far wall? _____

**Check your answers on page 151.**

## MULTIPLE STEP DIAGRAMS

Sometimes diagrams are used with a set of directions that tell how to do something. Here are the steps needed to fillet a fish (to cut the flesh from the bone).

1. Lay the whole fish on its side, holding it steady with one hand.
2. Using a sharp fillet knife, cut from head to tail to expose the backbone.
3. Cut the flesh crosswise, directly behind the gill. Then make a parallel cut along the ribs to the tail. Lift the flesh from the backbone.
4. Cut just behind the gill on the other side of the fish. Hold up the exposed backbone at the center and make a parallel cut along the ribs to the tail. Cut off the tail.

Notice how much clearer the directions become when they are accompanied by a diagram illustrating each step in the process, as in Figure 12.4.

## WARMUP B

**Use the diagram in Figure 12.4 to answer the questions.**

1. What is the title of the diagram? _____

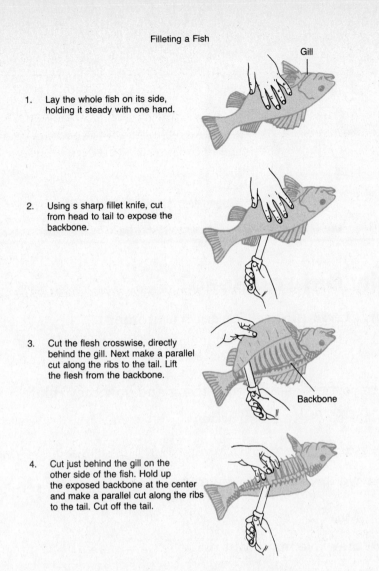

Filleting a Fish

1. Lay the whole fish on its side, holding it steady with one hand.

2. Using s sharp fillet knife, cut from head to tail to expose the backbone.

3. Cut the flesh crosswise, directly behind the gill. Next make a parallel cut along the ribs to the tail. Lift the flesh from the backbone.

4. Cut just behind the gill on the other side of the fish. Hold up the exposed backbone at the center and make a parallel cut along the ribs to the tail. Cut off the tail.

**Figure 12.4 A Multiple Set Diagram**

2. How many steps in filleting a fish does the diagram show? _____

3. When filleting a fish, what should you do after cutting the fish from head to tail to

expose the backbone? _____

4. During which step in the process do you cut off the tail of the fish? _____

5. Why are the directions for filleting a fish clearer when they are accompanied by a

diagram?_____

_____

**Check your answers on page 152.**

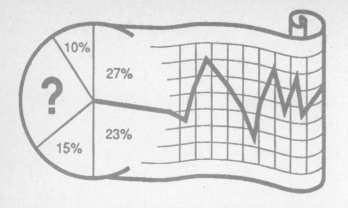

## How Carefully Did You Read?

**A. Choose the correct completion for each statement.**

1. Diagrams

    ☐ a. name the important parts of something and how they relate.

    ☐ b. take the place of written instructions.

    ☐ c. are often used as time lines.

2. A diagram will have a key when the diagram

    ☐ a. is drawn in color.

    ☐ b. shows more than five important parts.

    ☐ c. labels important parts with letters or numbers.

3. A likely subject for a diagram would be

    ☐ a. a plan of a backyard garden.

    ☐ b. a series of photographs of animals.

    ☐ c. the dates of birth of English kings and queens.

4. Diagrams are also useful in showing

    ☐ a. population numbers.

    ☐ b. the steps in building something.

    ☐ c. baseball statistics.

5. When reading a diagram, you should first

☐ a. study the key.

☐ b. study all the parts.

☐ c. look for the title.

6. Diagrams can be useful in

☐ a. clarifying written descriptions.

☐ b. showing trends.

☐ c. illustrating statistics.

**B. Use the diagram in Figure 12.5 to complete the statements.**

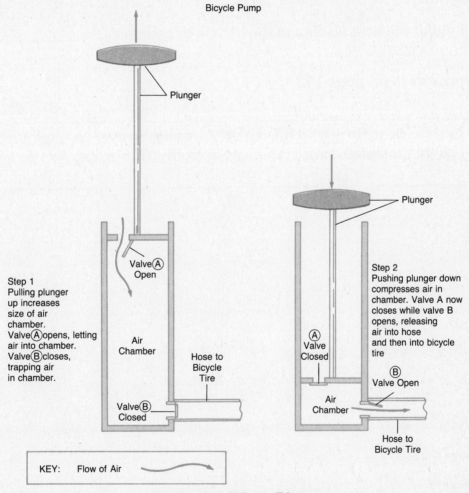

Figure 12.5  Using a Diagram

1. The diagram in Figure 12.5 shows _____

_____.

2. Step 1 in the diagram shows _____

_____.

3. Step 2 in the diagram shows _____

_____.

4. When the plunger is pulled up, the size of the air chamber in the pump _____.

5. Whenever valve Ⓐ is open, valve Ⓑ will be _____.

6. The key shows that ⌒➤ stands for _____.

7. In order to compress air in the air chamber, the plunger must be _____.

8. Compressed air gets to the bicycle tire when valve _____ opens, releasing air into the hose.

9. Air is kept out of the hose leading to the bicycle tire when valve _____.

**Check your answers on page 152.**

Look back at **How Much Do You Already Know?** on page 126. Did you complete each statement correctly? If not, can you do so now?

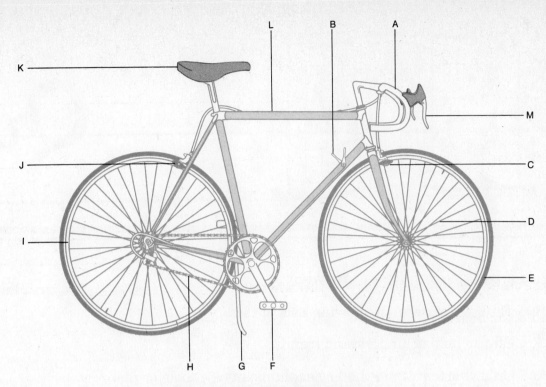

**Figure 12.6  Completing a Diagram**

ARE YOU READY FOR THE CHALLENGE of completing a diagram from information given you? If you think you are, write the name of each part near the appropriate letter in Figure 12.6.

Choose from these words: Chain, brake levers, frame, front brake, front wheel, gearshift, handlebars, kickstand, pedal, rear brake, rear wheel, seat, spokes.

Remember to put a title above the diagram.

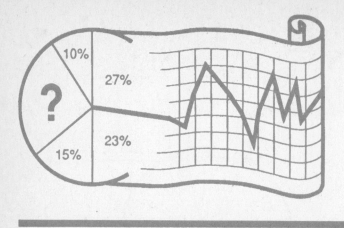

# Quiz

**A. Put an X beside each true statement.**

_____ 1. Tables are useful in showing lists of facts.

_____ 2. Charts help us understand trends.

_____ 3. On a chart or graph the key explains any symbols or pictures.

_____ 4. Some bar graphs are drawn horizontally instead of vertically.

_____ 5. A line graph is more accurate than a bar graph.

_____ 6. Combined bar and line graphs show the same information in two different ways.

_____ 7. Circle graphs are used to show parts of a whole.

_____ 8. Pictographs are as accurate as any other kind of graph.

_____ 9. Drawing a conclusion means guessing what the information on a graph really means.

_____ 10. A rising line on a graph shows an upward trend.

_____ 11. Time lines show when events happened.

_____ 12. Like time lines, flowcharts arrange information in a time sequence.

_____ 13. The first step in reading a diagram is to learn what the diagram shows.

# Understanding Tables, Charts, and Graphs

**B. Match each term with its definition. Write the letter of the definition in the space beside the term it defines.**

_____ 1. source

_____ 2. diagram

_____ 3. flowchart

_____ 4. pie chart

_____ 5. conclusion

_____ 6. symbol

_____ 7. trend

_____ 8. tipped graph

a. circle graph

b. object that stands for something else.

c. where the information on a table, chart, or graph comes from

d. upward or downward movement over a period of time

e. visual aid that shows steps in a process

f. graph that offers a way to compare two sets of figures

g. graph whose information may be distorted

h. visual aid showing parts of something and how they relate

i. a judgment based on facts

**C. Circle the answer that best completes each statement.**

1. To learn how accurate and reliable the information on a graph is, you should

   ☐ look for the source.

   ☐ study the scales.

   ☐ find the title.

2. Rows in a table are read from

   ☐ top to bottom.

   ☐ bottom to top.

   ☐ left to right.

3. On a chart or graph the symbol 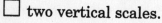 would likely be found in the

☐ title.

☐ source.

☐ key.

4. Bar graphs will always have

☐ the best information.

☐ a source.

☐ two scales.

5. Multiple line graphs are useful in

☐ pointing out errors.

☐ comparing two sets of figures.

☐ repeating the same information.

6. A combined bar and line graph is likely to have

☐ two vertical scales.

☐ two horizontal scales.

☐ no scales.

7. On a circle graph that pictures sections as percents, all the sections will add up to

☐ 100%.

☐ 1000%.

☐ 10.00%.

8. If the symbol ✈ stands for 1,000 airplanes, then ✈ will stand for

☐ 2,000 airplanes.

☐ 1,000 airplanes

☐ 500 airplanes.

9. An event that took place in the fifth century A.D. happened between the years

☐ 501–600.

☐ 401–500.

☐ 601–700.

10. On a flowchart the symbol ◊ stands for a

☐ start or end block.

☐ decision block.

☐ operation block.

11. A floor plan diagram shows

☐ what rooms in a building are made of.

☐ how well a building has been constructed.

☐ where things in a room or building are located.

**D. Use the graphs in Figures 1, 2, 3, and 4 to answer the questions.**

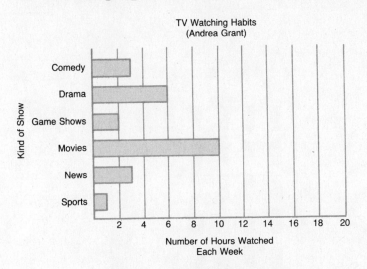

**Figure 1  TV Watching Habits**

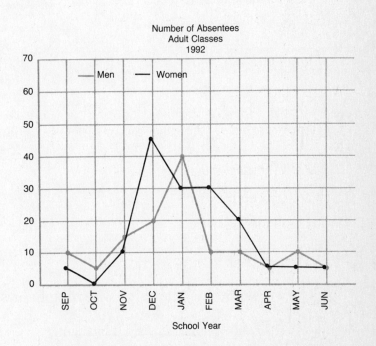

**Figure 2  Number of Absentees in Adult Classes**

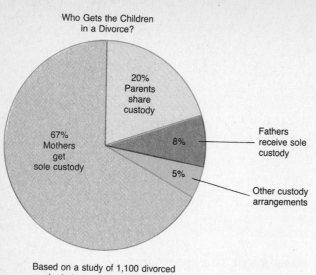

Who Gets the Children in a Divorce?

20% Parents share custody

67% Mothers get sole custody

8% Fathers receive sole custody

5% Other custody arrangements

Based on a study of 1,100 divorced couples between the years 1985–1990

**Figure 3  Who Gets the Children In a Divorce**

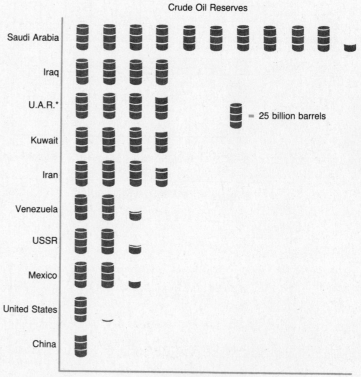

Crude Oil Reserves

Saudi Arabia
Iraq
U.A.R.*
Kuwait
Iran
Venezuela
USSR
Mexico
United States
China

= 25 billion barrels

*United Arab Emirates

**Figure 4  Crude Oil Reserves**

1. Figure 1 shows the television watching habits of Andrea Grant. What is her favorite kind of TV show? _____

2. How many hours of TV does Ms. Grant watch each week? _____

3. How many more hours of drama than sports shows does Ms. Grant watch in a week?

_____

4. How many hours of comedy shows does Ms. Grant watch each week? _____

5. What kind of TV show does Ms. Grant watch two hours each week? _____

6. What kind of graph is shown in Figure 2? _____

7. In what month were 45 women absent from adult classes? _____

8. What was the greatest number of men absent in any one month? _____

9. Women had perfect attendance one month of the school year. What was that month?

_____

10. How many men were absent in September? _____

11. What is the subject of the circle graph in Figure 3? _____

12. According to the graph, what percent of fathers receive sole custody of the children

after a divorce? _____

13. How many couples were involved in the study that resulted in the graph?

_____

14. What percent of divorced parents share custody of the children? _____

15. Of the total number of couples who were part of the study, how many were mothers

who got sole custody of the children? _____

16. Where did the information in the graph shown in Figure 4 come from? _____

17. What kind of graph is shown in Figure 4? _____

_____

18. How many billion barrels of oil does each ▮ represent? _____

19. How much oil does the United States have in reserve? _____

20. The country with the most oil reserves has how many more barrels of oil than the

country with the least reserves? _____

**Check your answers on page 152.**

# Do You Need to Review?

The following chart shows the numbers of all the items on the Quiz. You can use it to find which chapter you should review for any item you missed.

|            | Part A | Part B | Part C | Part D             |
|------------|--------|--------|--------|--------------------|
| Chapter 1  | 1      | 1      | 1, 2   |                    |
| Chapter 2  | 2, 3   |        | 3      |                    |
| Chapter 3  | 4      | 7      | 4      | 1, 2, 3, 4, 5      |
| Chapter 4  | 5, 10  |        | 5      | 6, 7, 8, 9, 10     |
| Chapter 5  | 6      |        | 6      |                    |
| Chapter 6  | 7      | 4      | 7      | 11, 12, 13, 14, 15 |
| Chapter 7  | 8      | 6      | 8      | 16, 17, 18, 19, 20 |
| Chapter 8  | 9      | 5      |        |                    |
| Chapter 9  |        | 8      |        |                    |
| Chapter 10 | 11     |        | 9      |                    |
| Chapter 11 | 12     | 3      | 10     |                    |
| Chapter 12 | 13     | 2      | 11     |                    |

# Glossary

The numbers in parentheses indicate the chapter in which each term is first used in this book.

**A.D.** (10) Abbreviation for the Latin term *Anno Domini* ("in the year of our Lord"); used for dates after the birth of Christ

**bar graph** (3) Graph that uses thick lines or bars to compare sets of figures

**B.C.** (10) Abbreviation for "before Christ;" used for dates before the birth of Christ

**cell** (2) The section of a table or chart where a column crosses a row

**century** (10) Measure of time equal to 100 years

**chart** (2) Information arranged in some orderly fashion, sometimes shown as symbols or pictures

**circle graph** (6) Graph in the form of a circle that shows parts of a whole, either as percents or amounts (See PIE CHART)

**column** (1) Vertical list of numbers or symbols on a table or chart

**combined bar and line graph** (5) Graph that includes both bars and one or more lines

**diagram** (12) Visual aid showing the important parts of something and how they relate

**extended bar graph** (3) Bar graph in which information given in a single bar is broken down into two or more smaller bars (See STACKED BAR GRAPH)

**flowchart** (11) Visual aid used to show the steps in a process

**graph** (3) Line, bar, or picture showing how one set of figures compares with another

**horizontal bar graph** (3) Bar graph in which the bars run from side to side

**horizontal scale** (3) Scale that runs across the bottom of a graph. Horizontal scales almost always measure units of time

**key** (2) Area of a graph that explains any pictures or symbols used on the graph

**line graph** (4) Graph that uses one or more lines to compare sets of figures

**multiple bar graph** (3) Bar graph in which more than one bar appears in the same scale

**multiple line graph** (4) Line graph that uses two or more lines to compare different sets of figures

**pictograph** (7) Graph that uses pictures or symbols to present statistical information

**pie chart** (6) Another name for a circle graph

**row** (1) Horizontal list of numbers or symbols on a table or chart

**scale** (3) Series of marks made along a line; used for measuring

**source** (1) Line at the bottom of a table, chart, etc., telling where the information came from

**stacked bar graph** (3) Another name for an extended bar graph

**symbol** (2) Object that stands for something else

**table** (1) List of information, usually presented as figures, arranged in some orderly fashion, especially in columns and rows

**time line** (10) Line that shows the order in which a number of events happened

**trend** (3) Upward or downward movement over a period of time

**vertical bar graph** (3) Bar graph in which the bars run up and down

**vertical scale** (3) scale that runs up and down on a graph

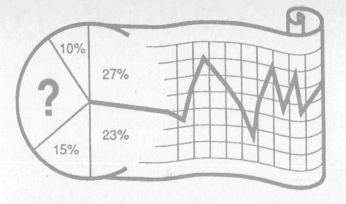

# Answer Key

## Chapter 1  Tables

### How Much Do You Already Know? (p. 2)

1. b     3. b
2. a     4. a

### Warmup (p. 5)

1. 212
2. Friday
3. Wednesday of Week 2
4. 1,036
5. 216 − 196 = 20

### How Carefully Did You Read? (p. 7)

**A.**

1. c     4. c
2. b     5. a
3. a     6. a

**B.**

1. Sales of Single Family Homes in Selected Rhode Island Cities
2. 310
3. 2,831
4. Middletown
5. Newport
6. 1986
7. Rhode Island Board of Realtors
8. 222

### Are You Ready for the Challenge (p. 9)

Title: U.S. Population 1790–1860

| Year | Population |
|------|------------|
| 1790 | 3,900,000 |
| 1800 | 5,300,000 |
| 1810 | 7,200,000 |
| 1820 | 9,600,000 |
| 1830 | 12,800,000 |
| 1840 | 17,000,000 |
| 1850 | 23,000,000 |
| 1860 | 31,000,000 |

## Chapter 2  Charts

### How Much Do You Already Know? (p. 10)

1. b     3. b
2. c     4. a

### Warmup A (p. 13)

1. 4
2. 124 (62 × 2 = 124)
3. 34 (17 × 2 = 34)
4. April–June
5. 12

### Warmup B (p. 14)

1. 32 miles
2. 313 miles
3. 607 miles (289 + 318 = 607)
4. 399 miles (461 − 62 = 399)
5. Elko and Boulder City (It is 481 miles between the two.)

### How Carefully Did You Read? (p. 15)

**A.**

1. c     4. b
2. a     5. b
3. c     6. a

**B.**

1. October Work Schedule: John Babyak
2. 21
3. 10
4. 5
5. 10
6. Sunday
7. October 30
8. Thursday

### Are You Ready for the Challenge? (p. 19)

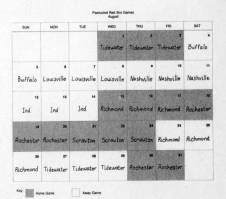

## How Much Do You Already Know? (p. 20)

**1.** c      **3.** b
**2.** b      **4.** b

## Warmup A (p. 23)

**1.** 10
**2.** 20
**3.** 10
**4.** 40 (50 − 10 = 40)
**5.** increasing

## Warmup B (p. 25)

**1.** beef
**2.** lamb
**3.** 58 lbs.
**4.** 72 lbs. (59 + 13 = 72)
**5.** 44 lbs. (59 − 15 = 44)

## Warmup C (p. 27)

**1.** 20
**2.** 5
**3.** 50
**4.** 1990 (25 men, 25 women)
**5.** 35 (85 − 50 = 35)

## Warmup D (p. 30)

**1.** 50 million
**2.** 51 million
**3.** 101 million
**4.** 187 million (138 + 49 = 187)
**5.** 113 million (161 − 48 = 113)

## How Carefully Did You Read? (p. 31)

**A.**

**1.** a      **5.** a
**2.** c      **6.** c
**3.** a      **7.** c
**4.** c      **8.** b

**B.**

**1.** multiple
**2.** cost of a 30-second ad on local TV news at 6:00 P.M.
**3.** $1,000

**4.** $650
**5.** $550
**6.** $1,900 (1,000 + 600 + 300 = 1,900)
**7.** $2,050 (800 + 700 + 550 = 2,050)
**8.** upward
**9.** $300 (950 − 650 = 300)
**10.** downward

## Are You Ready for the Challenge? (p. 35)

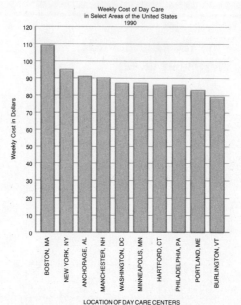

Weekly Cost of Day Care in Select Areas of the United States 1990

## Chapter 4 Line Graphs

## How Much Do You Already Know? (p. 36)

**1.** b      **3.** a
**2.** c      **4.** c

## Warmup A (p. 39)

**1.** 5
**2.** 30
**3.** May and September
**4.** 15 days (20 − 5 = 15)
**5.** a downward trend

## Warmup B (p. 41)

**1.** 1,000
**2.** 100 (1,000 − 900 = 100)
**3.** March (1,400 ads); June (1,450 ads); September (1,350 ads)

**4.** 300 (1,250 − 950 = 300)
**5.** 2,250 (1,050 + 1,200 = 2,250)

## Warmup C (p. 43)

**1.** 1,000
**2.** 1,350
**3.** 350 (1,350 − 1,000 = 350)
**4.** March 1991 (an increase of 600 ads over February 1991)
**5.** More ads in 1992

## How Carefully Did You Read? (p. 44)

**A.**

**1.** a      **5.** c
**2.** b      **6.** c
**3.** a      **7.** c
**4.** b      **8.** c

**B.**

**1.** multiple line
**2.** income and expenses for the Two Left Feet Shoe Company from 1988 to 1992
**3.** $110,000
**4.** $85,000
**5.** 1991 ($160,000)
**6.** 1988 ($85,000)
**7.** 1991 and 1992
**8.** $625,000 (85,000 + 110,000 + 125,000 + 160,000 + 145,000 = 625,000)
**9.** $620,000 (80,000 + 85,000 + 120,000 + 175,000 + 160,000 = 620,000)
**10.** go out of business

## Are You Ready for the Challenge? (p. 47)

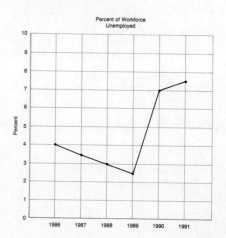

Percent of Workforce Unemployed

## Chapter 5 Combined Bar and Line Graphs

### How Much Do You Already Know? (p. 48)

1. c
3. c
2. b
4. b

### Warmup (p. 52)

1. 90°F
2. less than 1 inch
3. January (45°F)
4. June (82°F, no precipitation)
5. 45°F (90 − 45 = 45)

### How Carefully Did You Read? (p. 53)

A.

1. b
4. c
2. b
5. c
3. c
6. a

B.

1. 90° (June)
2. 72° (December)
3. 80°
4. February (about ¼ inch)
5. November (about 12½ inches)
6. 12¼ inches (12½ − ¼ = 12²⁄₄ − ¼ = 12¼ inches)
7. March
8. 52¼ inches
9. 7 inches (10 − 3 = 7)
10. Temperatures are too warm all year to allow snow.

### Are You Ready for the Challenge? (p. 57)

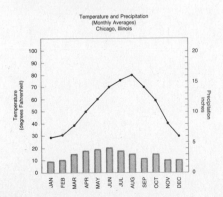

## Chapter 6 Circle Graphs

### How Much Do You Already Know? (p. 58)

1. c
3. a
2. a
4. c

### Warmup A (p. 61)

1. 36%
2. 46%
3. Unaccompanied Youth (4%)
4. 10% (14 − 4 = 10)
5. 100% (46 + 36 + 14 + 4 = 100)

### Warmup B (p. 62)

1. 29¢
2. Local Aid (31¢)
3. 30¢ (15 + 15 = 30¢)
4. 31¢
5. $1.00 (31 + 29 + 15 + 15 + 5 + 5 = 1.00)

### Warmup C (p. 64)

1. 12%
2. $240 ($2,000 × .12 = $240)
3. vacation
4. $1,200 ($2,000 × .60 = $1,200)
5. $200 ($2,000 × .15 = $300 for cable TV; $2,000 × .05 = $100 for books, magazines, and CDs; $300 − $100 = $200)

### How Carefully Did You Read? (p. 65)

A.

1. b
2. b
3. a
4. b
5. c
6. a ($4,000 × .12 = $480)

B.

1. The Federal Budget Dollar
2. where the dollar comes from
3. where the dollar goes (how it is spent)
4. 11¢

5. individual income taxes
6. excise taxes
7. 5¢
8. 32¢ (43 − 11 = 32)
9. 14¢
10. 25¢
11. 43¢ (22 + 21 = 43)
12. 2¢ (14 − 12 = 2)
13. 31¢ (25 + 6 = 31)
14. 6% (6¢ = 6/100 or 6%)

### Are You Ready for the Challenge? (p. 69)

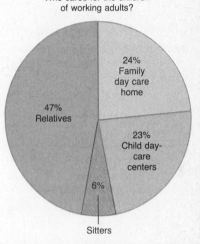

Who cares for the children of working adults?

24% Family day care home

47% Relatives

23% Child day-care centers

6% Sitters

Source: Office of Management and Budget.

## Chapter 7 Pictographs

### How Much Do You Already Know? (p. 70)

1. b
3. 5
2. c
4. c

### Warmup (p. 75)

1. 11 million
2. Italy and Spain
3. 24 million
4. 15½ million
5. 8½ million more (24,000,000 − 15,500,000 = 8,500,000)

### How Carefully Did You Read? (p. 76)

A.

1. c
4. b
2. c
5. c
3. b
6. a

**B.**

1. the number of American hunters by age groups
2. U.S. Fish and Wildlife Service
3. one million hunters
4. 25 to 44 years
5. 65 and older
6. 8 million (8,500,000 − 500,000 = 8,000,000)
7. 18 to 24 years
8. 500,000
9. 17.5 million
10. 4 million (8,500,000 − 4,500,000 = 4,000,000)
11. the 25- to 44-year-old age group, which is the largest group of hunters

### Are You Ready for the Challenge? (p. 79)

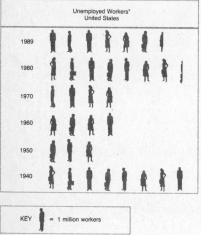

### Chapter 8   Analyzing Graphs

### How Much Do You Already Know? (p. 80)

1. a
2. b
3. c
4. c

### Warmup A (p. 83)

1. May, June, July, August, September
2. May and September (40°F)
3. 110°F (from − 45 to + 65 = 110)
4. 80° (from + 40 to − 40 = 80)
5. hard, since the weather is below freezing 7 months of the year

### Warmup B (p. 84)

1. 1940 (about 270,000)
2. 1945 (about 8.3 million)
3. 8,030,000 (8,300,000 − 270,000 = 8,030,000)
4. 1965
5. 1945, 1970

### Warmup C (p. 86)

1. Ethiopia ($121)
2. $12,060 (15,030 − 2,970 = 12,060)
3. 83%
4. Ethiopia
5. High per capita income and high literacy rates seem to occur in the same countries. (However, this does not mean that one causes the other.)

### How Carefully Did You Read? (p. 87)

**A.**

1. a
2. c
3. c
4. c
5. b

**B.**

1. oil use in the U.S. in 1970
2. oil use in the U.S. in 1989
3. 53%
4. industrial uses
5. 9% (15 − 6 = 9)
6. transportation
7. electric utilities
8. 12 % (4 + 8 = 12)
9. electric utilities, residential/commercial, and industrial
10. transportation

### Are You Ready for the Challenge? (p. 89)

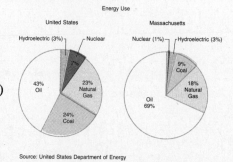

Source: United States Department of Energy

### Chapter 9   Seeing Is [Not Always] Believing

### How Much Do You Already Know? (p. 90)

1. c
2. b
3. a
4. b

### How Carefully Did You Read? (p. 98)

**A.**

1. c
2. c
3. a
4. c
5. a
6. c (400 airplanes)

**B.**

1. Couch Furniture Company
2. Ring Jewelry Company
3. $1,500
4. $4,000
5. 1985
6. $6,000
7. $1,000 in 1990
8. $2,000
9. $6,000 in 1985
10. $7,500 (17,000 − 9,500 = 7,500)

### Are You Ready for the Challenge? (p. 101)

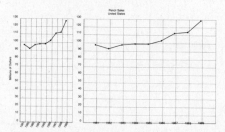

## How Much Do You Already Know? (p. 102)

1. b
2. c
3. c
4. a

## Warmup A (p. 105)

1. four
2. Babe Pratt
3. Maurice Richard
4. three (Toronto, Montreal, Chicago)
5. 1946

## Warmup B (p. 106)

1. five
2. Cabot visits Newfoundland (1497)
3. colony named New France (1663)
4. 1642
5. Cartier discovers the Gulf of St. Lawrence

## Warmup C (p. 108)

1. 3,000 years
2. 400 A.D.
3. Rome was founded
4. 500 years (1200 B.C. – 700 B.C.)
5. The Vikings began raids on what is now North America

## Warmup D (p. 110)

1. 1501–1600
2. 1001–1100
3. eighteenth
4. Texas broke away from Mexico (1835)
5. the sixth century B.C. (600 B.C. – 501 B.C.)

## How Carefully Did You Read? (p. 111)

### A.

1. b
2. b
3. b
5. b
6. b
7. c

4. b

8. c

### B.

1. The dates certain states were admitted to the Union
2. the *World Almanac and Book of Facts*
3. 1800–1900
4. seven
5. Alabama (1819)
6. 1850
7. Florida
8. Idaho and Wyoming
9. Alabama
10. Montana
11. seventy years (1889 – 1819 = 70)
12. twenty-five years (1845 – 1820 = 25)

## Are You Ready for the Challenge? (p. 115)

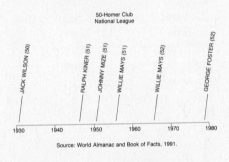

50-Homer Club
National League

JACK WILSON (50) | RALPH KINER (51) | JOHNNY MIZE (51) | WILLIE MAYS (51) | WILLIE MAYS (52) | GEORGE FOSTER (52)

1930  1940  1950  1960  1970  1980

Source: World Almanac and Book of Facts, 1991.

## Chapter 11  Flowcharts

## How Much Do You Already Know? (p. 116)

1. b
2. a
3. b
4. c

## Warmup (p. 121)

1. ten (including the steps involved in testing to see if the cake is done)
2. beat ingredients
3. remove cake from oven
4. three
5. test the cake for doneness

## How Carefully Did You Read? (p. 122)

### A.

1. c
2. c
3. b
4. a
5. a
6. c

### B.

1. hearing the alarm ring
2. Do I have time to sleep more?
3. reset the alarm
4. shower and get dressed
5. Do I have time to eat?
6. leave the house
7. seven
8. eleven

## Are You Ready for the Challenge? (p. 125)

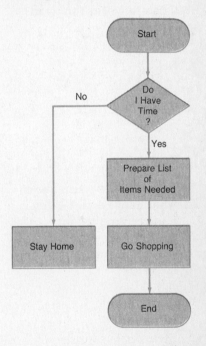

## Chapter 12  Diagrams

## How Much Do You Already Know? (p. 126)

1. c
2. a
3. c
4. c

## Warmup A (p. 132)

1. Fiction stacks
2. L
3. Reference stacks

4. M
5. A book by Isaac Asimov would be found in the fiction stacks to the left of the room. Alphabetizing books, from A to Z goes from left to right on the shelves.

## Warmup B (p. 132)

1. Filleting a Fish
2. four steps
3. Cut the flesh crosswise, directly behind the gill.
4. Step 4
5. The names of key parts of the fish are clearly shown in the diagram. It also shows how the knife is held.

## How Carefully Did You Read? (p. 134)

### A.

| | |
|---|---|
| 1. a | 4. b |
| 2. c | 5. c |
| 3. a | 6. a |

### B.

1. how a bicycle pump works
2. what happens when the plunger is pulled up
3. what happens when the plunger is pushed down
4. increases
5. closed
6. the flow of air
7. pushed down
8. Ⓑ

9. Ⓑ is closed

## Are You Ready for the Challenge? (p. 137)

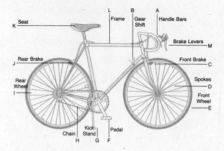

## QUIZ

### Understanding Tables, Charts, and Graphs (p. 138)

### A.

| | |
|---|---|
| 1. true | 8. false |
| 2. false | 9. false |
| 3. true | 10. true |
| 4. true | 11. true |
| 5. false | 12. true |
| 6. false | 13. true |
| 7. true | |

### B.

1. source (c)
2. diagram (h)
3. flowchart (e)
4. pie chart (a)
5. conclusion (i)
6. symbol (b)
7. trend (d)
8. tipped graph (g)

### C.

1. look for the source
2. left to right
3. key
4. two scales
5. comparing two sets of figures
6. two vertical scales
7. 100%
8. 500 airplanes
9. 401–500
10. decision block
11. where things in a room or building are located

### D.

1. movies
2. 25 hours
3. 5 more hours (6 − 1 = 5)
4. 3 hours
5. game shows
6. multiple line graph
7. December
8. 40 (in January)
9. October
10. 10
11. Who Gets the Children in a Divorce?
12. 8%
13. 1,100 couples
14. 20%
15. 737 (1,100 × .67 = 737)
16. Energy Information Administration
17. pictograph
18. 25 billion barrels of oil
19. 25.9 billion barrels
20. 231 billion barrels (255 − 24 = 231)